CODESIGN FOR
REAL-TIME VIDEO APPLICATIONS

CODESIGN FOR
REAL-TIME VIDEO APPLICATIONS

by

JÖRG WILBERG
GMD-SET

SPRINGER-SCIENCE+BUSINESS MEDIA, B.V.

A C.I.P. Catalogue record for this book is available from the Library of Congress

ISBN 978-1-4613-7786-3   ISBN 978-1-4615-6081-4 (eBook)
DOI 10.1007/978-1-4615-6081-4

Printed on acid-free paper

Contents

List of Figures

List of Tables

Acknowledgments

The book is a summary of a PhD thesis from the *Brandenburgische Technical University of Cottbus*, GERMANY.

The thesis was prepared in the Sydis / CASTLE project at GMD-SET. The work was sponsored by the BMBF (Bundesministerium für Bildung, Wissenschaft, Forschung und Technologie) under research contract 01M2897A Sydis.

I would like to thank my supervisors (in alphabetical order): Prof. Dr. R. Camposano, Prof. Dr. W. Rosenstiel, and Prof. Dr. H.-T. Vierhaus for all the support and many helpful discussions.

This work would have been impossible without the support and friendship of the Sydis / CASTLE team (in alphabetical order): S. Gehlen, Dr. A. Gnedina, Dr. H. Guenther, H.-U. Kobialka, A. Kuth, Dr. M. Langevin, G. Pfeiffer, P. Plöger, J. Schaaf, U. Steinhausen, Dr. P. Stravers, M. Theißinger, Dr. H. Veit, U. Westerholz.

Thanks to all the people at GMD-SET for providing such a splendid working environment.

I would like to acknowledge the support from the MOVE group at the Delft Technical University, in particular, Dr. H. Corporaal and Dr. J. Hoogerbrugge. GNU and other public domain software was used extensively for the program development and the preparation of the thesis. Thanks to all the people who contributed to these good programs. The text of the thesis was prepared with the GNU EMACS editor and the *auctex* package. The typesetting was done with the LATEX system. The figures were prepared with *gnuplot* and *xfig*.

Last but not least, I would like to thank my friends and family for all the encouragement and sympathy that helped me throughout the preparation of the thesis.

1   INTRODUCTION

Design automation is a service to the designer in the first place [131]. It facilitates the design of increasingly complex systems in less time. This is typically achieved by raising the level of abstraction at which the design is specified [133]. Advancing to the system-level requires new design methodologies. They have to be developed along realistic application examples, or as Minsky and Papert [138] have put it:

> *"Good theories rarely develop outside the context of a background of well-understood real problems and special cases. Without such a foundation, one gets either the vacuous generality of a theory with more definitions than theorems - or a mathematically elegant theory with no application to reality. Accordingly, our best course would seem to be to strive for a very thorough understanding of well-chosen particular situations in which these concepts are involved."*

This thesis investigates the design of real-time video processing systems. They reduce the data rate for digital video signals. It is a key technology for transmitting video data across computer networks or for storing them in

1

databases. Making video compression a central element of multimedia technology. Multimedia combines different media types (audio, text, video) with computer programs (database searches, hypertext navigators, information agents). Video data are especially important in this context, because they make the human-machine interface more natural. Vision is our most dominant sense. Information can be presented in an attractive way. Video compression is essential at this point. It allows to store and exchange multimedia documents without excessive bandwidth consumption [146]. Standards are necessary to make the compression independent from specific platforms or vendors. The best compression is often achieved by exploiting specific application characteristics, for example, a person sitting in front of a video phone. A multimedia scenario (chapter 2) will typically combine several standards for different application scenarios. This means, the video compression system must provide flexibility in conjunction with high processing power.

Both can be achieved by an embedded video compression system. It is integrated into a host system, like a PC or a workstation. High processing power is achieved because the embedded system is streamlined for video processing. All the general functionality (like network interfacing or file management) is provided by the host system. The embedded system is focused on specific tasks. But it is not a dedicated system. It can be programmed for different compression standards. This programmability gives the flexibility that is of prime importance in the multimedia domain. Put in other words, the embedded system separates from a general-purpose processor because it addresses a narrow application domain. Specific characteristics are exploited to improve the performance. But it is programmable and therefore more flexible than an application-specific hardware solution.

The design of an embedded video compression system requires the codesign of hardware and software for a narrow application domain. Special purpose software is almost inevitably a part of an embedded system. Therefore the design and optimization becomes a major target of the design, thus demanding the codesign approach for embedded systems. A restricted form of codesign is the automatic partitioning [47] [67]. In this case a fixed hardware structure is assumed, consisting of a general-purpose processor and a dedicated ASIC. The partitioning forms hardware (HW) and software (SW) tasks. The hardware tasks are mapped onto the ASIC, while the software tasks are handled by the general-purpose processor. One of the main obstacles, in realizing the automatic partitioning, is the transformation of algorithms. To decide, whether a task is best suited for hardware or software, one must compare the most suited hardware formulation with the best possible software formulation. This means, the partitioning system must automatically create variants of algorithms. For real-life designs, these variants are completely different. A good hardware solution often has little in common with a good software solution (see the priority queue example in [78]). Knowledge about the application domain is required when creating the variants. This is not addressed in an all-automatic approach.

A more general form of codesign is the *combined* design of hardware and software. This is a well-established approach for RISC processor design [31] [76] [75]. It avoids an early partitioning of hardware and software. The design focuses on both domains concurrently. This allows to exploit synergetic effects, and the designer's knowledge about the application context can be used.

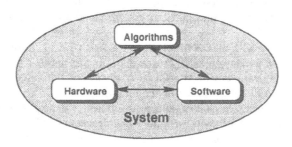

Figure 1.1.   Design domains that are combined in embedded system design. The designer can control hardware, software, and algorithms.

While this is already a successful technique in general-purpose processor design, the prospects are even better when looking at embedded system design. The designer concentrates on a narrow application domain. He/she controls the software, the hardware, and even the algorithms (see Fig. 1.1). The processor can be tailored for this application domain. The software is fine-tuned for the processor and even the algorithms can be selected so that they fit best to this hardware-software combination. The designer optimizes the system as an entity because he/she controls *all* design domains.

Unfortunately, the drawback of this approach is the complexity. A nearly infinite number of variants is possible. Implementing them in detail is prohibitive due to the required design time. An abstract approach is necessary. This is addressed in the *quantitative analysis* that forms the core of Hennessy and Patterson's "Quantitative Approach" [76]. The quantitative analysis provides engineering data for designing a system. Data on possible system trade-offs are measured. Design decision are based on these data. This avoids detailed implementations in an early phase of the design. But the approach cannot be used directly for embedded system design. The analysis data are missing. In general-purpose processor design, the data can be taken from literature, since the design is aimed at a broad application domain. Processor designs can share these data. In embedded system the situation is exactly the opposite. The special characteristics of a narrow application domain are essential to the design. The data must be collected for this domain. Therefore, analysis tools are necessary which perform measurements automatically. They provide a picture of system trade-offs on an abstract level. No variants are designed in detail. An abstract design becomes feasible. This abstract design must be translated into the final system implementation. Again, design tools are necessary. They create the hardware and software descriptions. The resulting system exploits the specific characteristics of a narrow application domain, thereby improving the performance. At the same time, flexibility is provided due to the programmability.

Realizing the approach means to solve three main tasks: (1) a stepwise design methodology must be developed, (2) design tools are necessary, and (3) the approach must be evaluated for a realistic application, the video compression in this case. This will be described in the following.

The remainder of the thesis is organized as follows. The next section presents related work in the field of synthesis and video processing. Chapter 2 summarizes the background material on video processing and describes the embedded video compression system. This is used as a design project throughout the thesis. An outline of the design methodology is given in chapter 3. One of the core parts is the quantitative analysis. A detailed discussion is presented in chapter 4. The structure of the design tools is explained in chapter 5. The overall organization and the result-documentation is performed by an HTML-based framework (chapter 6). Finally, chapter 7 exemplifies the approach by looking at the results of the video compression system.

MOTIVATION AND RELATED WORK

This section reviews some recent work in the field of synthesis and in the field of video processing. Due to the variety of work in both areas the overview is not meant to be exhaustive.

The synthesis approaches are usually classified by the level of abstraction at which the designer specifies the system [133] [28] [54]. High-level synthesis [28] [191] takes a behavioral specification as an input. This specification is typically written in an HDL (Hardware Description Language), like VHDL or Verilog. The specification is transformed into an intermediate representation, e.g., a CDFG (Control Data Flow Graph), from which a datapath and a control unit is generated. In most cases, the datapath is configured from RT (Register Transfer) level components. The controlling is hardwired into an FSM (Finite State Machine).

Recently high-level synthesis has been integrated into commercial design systems [32] [128] [173]. Examples of these tools are Synopsys' BC, Mentor's Mistral2, BooleDozer, Asyl+, etc. The current research work in the synthesis area tries to increase the level of abstraction. The goal is to support the design of complete systems [27] [136] [208]. This work can be subdivided according to the different topics that have to be addressed when designing systems [208].

An important aspect is the system **specification** and the formal **verification** of consistency within these specifications. Typically synchronous languages [18] [70] like Lustre [71], Esterel [24], and StateChart [73] or synchronous subsets of existing HDLs (e.g., VHDL in [15]) can be used. Some design tools use this specification paradigm. Examples include ASAR [11], Cosmos [93] [102], Codes [25]. ASAR defines a framework for codesign. The specification is written in Signal, Lustre, etc. A hardware/software partitioning is performed. Structural VHDL is generated for the hardware part and C code is generated for the software part. Cosmos translates specifications, written in different specification languages (SDL, StateCharts, etc.), into a common intermediate format called Solar. The internal description is partitioned into hardware and software parts, and mapped onto an architecture template. Codes adopts a PRAM model [33] as paradigm of the target architecture. SDL or Statemate is used for the specification. C code is generated for the software and the hardware is configured from off-the-shelf components. The approach for the hardware design is similar to MICON [21] [64]. MICON is an expert system that supports the configuration of computer systems. Standard components like microprocessors, RAMs, I/O devices etc. are used. The rules of the expert system select the component families that are suitable for a combination, and they define the type of connections that have to be used when creating the circuit. The approach allows to reuse design information from previous designs when working on a similar design project.

The **simulation** of hardware/software codesigns is addressed in Ptolemy [26] [106]. Ptolemy is written in C++ (user interfaces are written in Tcl/Tk). It offers a class library that encapsulates the communication among different simulation tools. A framework for building complex simulation environments is formed. These environments can be heterogeneous, consisting of different types of tools which are connected by the Ptolemy classes. The main advantage of this approach is the flexibility and the embracing of heterogeneity [26] [108]. No uniform design approach is required, rather a suitable design environment can be formed by connecting the tools appropriate for the current design project. It provides flexibility for the multitude of different application domains, which are typically encountered when designing embedded systems.

Automatic **partitioning** between hardware-tasks (i.e., tasks realized by dedicated hardware) and software-tasks (i.e., tasks mapped onto a general-purpose microprocessor) is addressed by many different projects on design automation. Examples include VULCAN [67] [66], Cosyma [47], Tosca [5], and the approaches by Kalavade [107], Olukotun, et al. [148] among others. The current version of these approaches is not suitable for designing high-performance systems because a very simple target architecture is often assumed for the partitioning. In most cases this target architecture consists of a small microprocessor without memory hierarchy. A dedicated ASIC acts as a kind of co-processor. The partitioning takes place on a fine grained level (i.e., on an instruction or basic-block level). In some systems the microprocessor and the dedicated hardware do not even operate in parallel. Todays microprocessors are much more complex [72] [159] [171] [194], they use a complex superscalar architecture [103] and rely heavily on memory hierarchy [162] [17] (even the L2 cache is integrated on the processor chip in [171]). This is the basis for achieving high performance. Therefore it is normally not possible with these processor, to perform certain instructions externally by some dedicated hardware (typically the performance lost due to the L1 cache miss and the accompanied data hazards will far outdo all performance improvements that might be possible with the dedicated external hardware). This means, fine grained parallelism should be exploited by instruction level parallelism [168] within the processor. But automatic partitioning on a coarse-grained level would require some kind of parallelizing compiler that automatically extracts parts of the specification which can be executed in parallel. This has been a research topic in multiprocessing since a long time [210] [158] [90] [77] [113]. But good performance is only achieved if the designer controls the parallelization [29] [110]. Thus extending the automatic partitioning approaches to complex, high-performance applications seems at least very difficult.

A **performance analysis** based on the pixie tool is used for the software part in [148]. In [60] a code generation for VLIW processors is used to estimate the performance improvements achieved by higher degrees of instruction level parallelism. In [84] an analysis determines the complexity of the main tasks in a powertrain module. In [100] an RT-level estimation technique is described for bounding the hardware complexity of pipelined modules. This technique is also used as a basis for the partitioning in the USC tools [65]. However the analysis is currently not well integrated into the design environment.

In the approaches mentioned so far, software tasks are handled by using an existing compiler of a general-purpose processor. This limits the design choices to existing standard processors. Recently **retargetable code-generation** has gained more attention. There are two main motivations for the research on retargetable compilers: in case of DSPs, the existing compilers are often not good enough, and in case of ASIPs (Application Specific Instruction Set Processors), the complexity makes manual programming a tedious and error prone task.

Mimola [132] and Capsys [10] generate firmware for VLIW processors. The Capsys system can directly process C code. The PEAS-1 [4] system uses a modification of machine descriptions in the GNU gcc to produce opcode for a newly designed processor. This approach also allows to process C code, but only a RISC-like scalar processor architecture is supported as target architecture. The FlexWare [124] system supports more general DSP architectures. A pattern matching is used to exploit the special instructions offered by these processors. The Cathedral Chess [160] [119] system uses a probabilistic approach for the code generation. The approach is similar to the force-directed scheduling [151] known from high-level synthesis. In [140], a BDD based exploration of the processor structure is used for the code generation.

A number of **case studies** have been performed to explore codesign in different application domains. The simulation capabilities of Ptolemy are exemplified by the simulation of a telephone transmission line [106]. A priority queue was used as an example in [78] to indicate the problems of automatic partitioning. A private telephone switch board was used as an example in [209]. A main problem in this context was to decide upon the most suitable implementation of the touch-tone detection. An analog device was manually selected in that case. It is difficult to imagine a design system that can automatically perform such decisions. Therefore it is important that the tools assist the designer. He/she can make the best decisions for the important parts of the design.

The design of a portable computer called VuMan is described in [178]. The design time was significantly reduced in comparison with the manual design of the first version. This was achieved by a concurrent engineering of hardware and software. Simulation and rapid prototyping was used extensively to avoid design errors as much as possible. A case study for automotive electronics is described in [84]. A performance analysis was used as a basis for the main design decisions. The importance of such an analysis is also underlined in [127] where the design of a graphics processor is investigated. The design of a dedicated JPEG video codec is used as an example for system-level synthesis in [65]. The main emphasis was put on a system-level partitioning. The results obtained in this case are comparable to the first JPEG codecs [12] [22] [164] which were specified at lower levels of abstraction.

The first generation of video compression chips can be subdivided into two groups: dedicated ASICs and homogeneous multiprocessors. Dedicated ASICs are used in a number of different approaches. The first JPEG chips were built in this way. Dedicated chips are also the core of the first H.261 codecs build by Bellcore [52], LSI-Logic [170], and Plessey [37] [38]. Another example of this approach is the first MPEG decoder developed by C-Cube [163]. In these cases each task of the algorithm is directly implemented by a dedicated hardware module. The main disadvantage of this approach is its inflexibility [

13]. For example, the LSI-Logic codec was designed with respect to the RM8 standard, a preliminary version of the H.261 standard [125]. A compliance with the new standard requires a redesign of the chips. A second problem of the dedicated architectures is the difficulty in matching the throughput of the various modules. Often these modules should be used for many different applications. This allows to produce them in higher volumes. An example thereof is a DCT (Discrete Cosine Transform) module. It is an integral component of several video compression standards (see chapter 2). Therefore, it is useful, to design a DCT module such that it can be used for many different compression applications. The applications can range from low bit rate applications like hand-held video phones to high performance applications such as HDTV. However, the modules must meet the throughput demands of the most demanding application, HDTV in this case. Thus the modules are typically oversized for all other applications. Unfortunately, they cannot be used for other tasks since they are dedicated modules. The result is an ill-balanced system that wastes resources.

The situation can be avoided by programmable systems. Therefore the second group of video processing systems used multiprocessor based solutions [53] [137] [143] [187] [212]. Typically different image segments are assigned to different processors. Each processor performs the complete compression algorithm for the given image segment. A data parallelism is exploited based on an SPMD (Single Program Multiple Data) approach. Since the video processing algorithms handle all image segments in a similar way, a good load balancing can be achieved. As a disadvantage, the special properties of the different compression tasks are not exploited [202] [197].

This can be achieved by heterogeneous multiprocessors. A combination of both aforementioned approaches is used. The system is constructed of several processors. Each of these processors is adapted to a certain range of tasks. This is the main difference to a dedicated module which can handle just a single task. In comparison to a homogeneous processor, the adapted processors achieve better performance because they exploit the specific characteristics of particular tasks. They can use special optimizations. Examples of these more recent approaches include C-Cube' s CL4000 [193], IIT' s VPC [14], AT&T' s AVP4000 [2], Matsushita' s VDSP [7], and TI' s MVP [49] [68] [69]. Most of these processors have specialized modules for tasks like blockmatching or the DCT (Discrete Cosine Transform) and programmable modules for the system controlling, managing of different data streams, etc. For example, AT&T' s AVP4000 uses a SPARC like RISC core and 45k lines of microcode for controlling the compression system. Another line of development is the integration of multimedia support into general-purpose microprocessors. This is a result of the extreme popularity of multimedia applications. It forces designers of general-purpose computers to pay more and more attention to this domain. Examples of this development include the 88110 [43], the PA [122] [121], the UltraSparc [215], Pentium-MMX [91]. The general strategy followed in all these cases is similar. It was pioneered in the IIT video processor.

Video processing often requires large throughput where many operations are performed on large streams of data. Frequently, these operations can be performed in parallel. Furthermore, the word width of the data (8-*bit* for pixels, 16-*bit* for intermediate results) is much smaller than the word width of conventional processors (currently a word width of 32-*bit* or 64-*bit* is used). The multimedia operations allow to split the addition (or other operation) into a number of sub-operations. For example, a 64-*bit* adder can be used to perform eight 8-*bit* operations in parallel. This alternative requires little additional hardware. In principal, only the carry chain of the adder must be broken. This is normally not a problem if a carry-lookahead or a carry-select adder is used. Thus the performance for multimedia applications can be increased without adding much hardware, as little as 0.2% in [122].

A second modification, which is often used to support multimedia, is the saturation arithmetic. Traditionally, microprocessors use a modulo arithmetic. It is often based on 2's complement representation of numbers. In this case an overflow of a positive number wraps around into the negative numbers. For example, adding a 1 to the largest positive integer will give the largest negative integer which can be represented by the word width of the processor. In video processing, the values of the numbers are often used to represent the grayscale values of a pixel. In the example, this would mean, that a white pixel is turned into a black pixel, the other extreme in the grayscale. Of course, this is not desirable. It will lead to sparkling noise in the images, especially when filter operations try to use the complete dynamic range of the grayscale.

Saturation arithmetic handles the overflow situation more efficiently. It performs a clipping to the largest value when an overflow occurs. The value is saturated and cannot be increased any more. In the example above, adding a value to a white pixel will leave the pixel unchanged. The white pixel is already at maximum brightness. This type of operations allows to use the full dynamic range in filter operations without excessive checks for overflows. No undesirable noise will be produced. This speeds up the performance especially in superscalar processors where branches often cause a significant delay.

A more comprehensive overview of the different video processors can be found in [111] [202] [197]. The next section gives an overview of the video compression system which is used to exemplify the codesign approach described in this thesis.

2 DESIGN PROJECT: VIDEO COMPRESSION SYSTEM

This section gives some background information on the video compression algorithms and it defines the main structure of the video compression system. These informations will be used in the remainder of the thesis to exemplify the main steps of the design flow.

Video compression is used to reduce the data rates for storing or transmitting video sequences. Set-top boxes for video-on-demand [155] and multimedia computer systems [61] [134] [88] are two important application fields. Video-on-demand will require the decompression of movies which are fed into a cable network. Little flexibility is required in this case since only a single application must be handled. Multimedia systems, on the other hand, require much more flexibility. Typically a variety of different applications are processed on the same system [50] [146]. These include video conferencing [125], multimedia database access [112] [19] [161], CD-I [79]. For example, a user of a multimedia system likes to create a World-Wide Web document that contains photos, video sequences, and normal text. The user might create this document in collaboration with a colleague at some remote site. They will use a video phone to communicate while editing the document. In this case, the multimedia system has to process several different video compression standards almost concur rently. The video phone will typically use H.261 for compressing the video data. The still pictures are compressed by JPEG and the video sequence might use MPEG to allow VCR like features when displaying the sequence. These standards will be described in more detail in section 2.1. The next section recalls the fundamental structure of a transmission line for compressed data. It defines the real-time condition. This condition governs the throughput that must be achieved by the embedded video compression system.

REAL-TIME CONDITION

The generic structure of a transmission system for compressed data is shown in Fig. 2.1. Two main tasks can be distinguished: the source coding and the channel coding. The source coding is concerned with the reduction of redundancy. Often symbols emitted by the information source are not independent from each other. For example, adjacent pixels in an image have the same color because they belong to an object, e.g., the sky in a holiday photo. In this case the symbols can be predicted from the context of the surrounding symbols. This means, if the pixel, which is currently encoded, belongs to the sky, it is very likely, that the next pixel also belongs to the sky. Thus it is very probable, that the new pixel will have the same color and brightness. It can be predicted from the current pixel. This kind of redundancy increases the amount of data. But it does not convey any further information to the receiver. Hence, transmission cost are wasted by sending unnecessary information. The source coding is used to avoid this wasteful situation. The source coder transforms the source symbols into a stream of code words. In the ideal case, the coded data would contain no more redundancy. They encode exactly the essential information. But this makes the data also vulnerable for any disruption. Any errors that occur during the transmission would immediately cause a loss of information. To prevent such a loss, an error correction code (ECC) is added. This is done by the channel coder. It creates a robust data package where changes of the data, which occur during the transmission, can be detected and corrected by the channel decoder. The output of the decoder will contain the data sequence that was produced by the source coder (at least in the ideal case). This sequence is decoded to reconstruct the original data.

Video compression is mainly concerned with the source coding and decoding strategies. This is the main focus of the compression algorithms addressed in this thesis. One of the fundamental constraints on the processing is the real-time condition. The compression and decompression must be done without causing noticeable delays in the image sequence. The compression system must be designed in such a way that it can handle all the different image sequences and compression algorithms in real-time. But the system should not be oversized. This would increase the costs and thus reduce the market competitiveness. The system should fit exactly for the considered application domain. This will be investigated in more detail in the following.

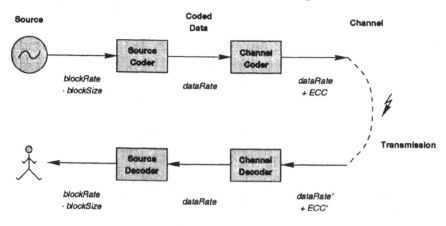

Figure 2.1. Basic transmission scenario for compressed data (ECC: Error Correcting Code).

The main figure of merit in the source coder is the achieved compression ratio. It is defined as follows

$$compression\,Ratio = \frac{blockRate \cdot blockSize}{dataRate}.$$

The stream of data from the source is subdivided into blocks. The term *block* is used as a generic term in this thesis. It denotes an element of fixed size from the video data stream. For most analysis data, a macroblock is used. It consists of 16×16 pixel for the gray scale (luminance) component and 8×8 pixel for each of the two color (chrominance) components. *blockSize* defines the size of these blocks in bit. The video compression standards of section 2.1 use 8-*bit* per pixel. This gives a macroblock size of 384*Byte*. For the remainder of the thesis it is assumed that the complexity of the source coding algorithms are almost linear in the number of processed blocks. The blocks are emitted at a constant rate (*blockRate*) by the source. The block rate depends on the size of the image frames and on the rate of these frames. Commonly used formats [144] [92] [157] are summarized in Table 2.1. The *blockRate* times the *blockSize* defines the number of data produced by the source per second. After applying the compression ratio, a lower *dataRate* is achieved. The *compressionRatio* defines the reduction of data volume after applying the source coding algorithms. It this is in the order of $10 \ldots 100$ for the compression standards (see Table 2.2).

Table 2.1. Commonly used image formats and number of macroblocks (16×16 pixel luminance and 8×8 pixel for the chrominances).

Format	Frame Size Luminance	Frame Rate	Macroblocks / Second
QCIF	176×144	29.97	2949.21
CIF	352×288	29.97	11796.84
SIF	352×240	29.97	9830.70
CCIR 601 *)	720×576	25	40500.00
HDTV (J) *)	1920×1035	30	232875.00

*) A vertical subsampling of the chrominance components by a factor of 2 is assumed.

The compression ratio is achieved at the cost of processing power which is provided by the compression system. The compression system must be fast enough to process the source coding algorithm for the current block, before the next block is emitted by the source (*real-time* condition). If the compression system processes the blocks much faster than they are emitted by the source, it has to insert idle cycles. This would be a waste of resources in the compression system. A smaller, less expensive system can be used for the same task. So the ideal compression system should be exactly as fast as necessary to process the blocks emitted by the source. This is the main task of the system design. It is complicated by the fact that a number of different source coding algorithms have to be considered. Furthermore, the processing time depends on the different types of video sequences and on the compression ratios. All these aspects must be carefully balanced in the embedded system design. This is the main topic addressed in the thesis. As a starting point, we define the real-time condition more precisely. It is convenient to model the source as a probability space $(\Omega, \mathbf{B}, \mu)$ [62]. Ω is the set of all image sequences. $\mathbf{B}$ is the σ-field of all possible subsets of Ω and μ is the probability that an image sequence $\omega \in \Omega$ is emitted by the source. The compression system associates a $\frac{time}{block} : \Omega \to \mathcal{R}_+$ with each block. The time depends on the coding algorithm, the software implementation, and the hardware structure of the compression system. Furthermore it depends on the contents of the image sequence that is coded. We define the set of image sequences (S) that achieve a processing time in a certain interval $I \subset \mathcal{R}_+$ as follows

$$S \equiv \frac{time}{block}^{-1}(I) \equiv \{\omega \in \Omega : \frac{time(\omega)}{block} \in I\} \ .$$

As an example we can assume the processing of video sequences in QCIF (Quarter CIF) format. There are approximately 2950 macroblocks in each second. This means a macroblock, has to be processed in about 0.33 milliseconds. If the video processing system needs more time to handle a macroblock, it cannot handle QCIF in real-time. In other words, a processing time in the interval $I = (0, \frac{1}{2950})$ will not violate the real-time condition. A processing time in the interval $I = [\frac{1}{2950}, \infty)$ is too slow for this application. An exception mechanism has to handle this situation. More formally, a block would violate the real-time requirement, if the processing of the block is not completed before the next block is emitted by the source, that is, if the processing time is in the interval

$$I_{except} \equiv \left[\frac{1}{blockRate} , \infty \right) \ .$$

The *real-time condition* can be defined as follows

$$\mu(S_{except}) \leq P_{except} , \tag{2.1}$$

$$where \quad S_{except} = \frac{time}{block}^{-1}(I_{except}) \ .$$

S_{except} is the set of all image blocks that cannot be processed in time. The probability of these sequences $\mu(S_{except})$ should be less than a selected exception probability. In most cases it is not necessary, that all sequences achieve the real-time condition. Often it might be permissible that some exceptional sequences can violate it, as long as, their occurrence probability is sufficiently low. Thus, the real-time condition can be further refined into *hard* real-time tasks, in this case P_{except} is zero, and *soft* real-time tasks where $P_{except} > 0$. Hard real-time tasks are typically encountered in safety critical applications like flight controllers or anti-block brake control systems. In this case a response of the system that exceeds a deadline might cause a major accident. Therefore the real-time condition should always be fulfilled. Video processing tasks often belong to the soft real-time tasks. In this case, a frame might be skipped if it cannot be processed in-time. This might be annoying for the user but it will not cause any severe damage. The designer must only guarantee that P_{except} is sufficiently small. The most suited value of P_{except} depends on the anticipated market segment. High-end users, like TV studios, might look for a system which will process almost every sequence, while users of low-cost systems, like PC or video games, might tolerate some quality degradation as a price for lower system costs. Equation 2.1 suggests two ways of proving that the real-time condition is fulfilled. First, we can make sure that the processing time is always smaller than the block period, i.e., $\frac{time}{block} \leq \frac{1}{blockRate}$. In this case S_{except} is empty so that even the hard real-time condition is fulfilled. A typical way to assure this is to use worst-case conditions for all data dependent control flow operations. This determines the longest possible path in the program. In addition, data dependent loops or unbounded recursion have to be avoided. Furthermore, the worst-case execution times have to be assumed for the hardware. For example, caching cannot be considered unless it can be guaranteed that all data are in the cache. Speculative execution and pipelining are also difficult since the execution time might be data dependent. This means, any kind of speculative execution must be avoided in the processor. But this is one of the key components for achieving high performance in modern processors [150] [179]. Disabling all these features means a severe performance degradation. Hence it should be used only for extremely critical tasks.

For video processing it is more attractive to show that the probability of violating sequences $\mu(S_{except})$ is sufficiently small. A mathematical proof is very difficult. It would require to know the occurrence probabilities of all image sequences and their respective processing times. A practical approach to use a typical ensemble of image sequences which reflects the properties of the set of all possible sequences. A measurement of the processing times is performed and the occurrence probability of the sequences is estimated. For example, a video coder, which is always used in conjunction with a low-bitrate channel, need not handle video data where the data rate after compression exceeds the available transmission capacity of the channel. This is an important restriction for designing wireless communication systems. In practice, an ensemble of typical sequences is selected. The selection has to ensure, that the sequences are representative for most image conditions.

Before describing the measurement and analysis (chapter 4) and the necessary design tools (chapter 5) in more detail, a description of the standard video compression algorithms will be given.

VIDEO COMPRESSION ALGORITHMS

The example at the beginning of this chapter showed two important requirements for video compression in the multimedia context: (1) different application areas (still picture compression, video phones) are combined, and (2) data must be exchanged on a platform independent base. Especially the exchange of documents via the WWW (World Wide Web) [19] or multimedia email [23] motivates the use of compression standards [172]. In this case the hardware and software of different vendors can be combined. The cost for developing multimedia applications are justified by the large number of users. Currently the DCT (Discrete Cosine Transform) [3] based standards [8] [56] [144] [172] like JPEG [154] [192] [95], H.261 [125] [97], and MPEG [55] [94] have gained broad acceptance. Although these standards are not the only possible compression algorithms, they are the most important once because they found such a widespread acceptance. Especially the generic standards like MPEG and JPEG are successful. They can be adapted to a variety of different application domains and many trade-offs between the system costs and the achieved performance are possible.

The algorithms exploit the redundancy encountered in a video sequence and transform the sequence into a representation which requires fewer bits during the transmission. The main forms of redundancy in a picture sequence are temporal and spatial redundancy. Additionally, the image quality can be gracefully degraded. This exploits the fact that the human vision system is very tolerant to certain changes. For example, the sharpness of the edges can be reduced without annoying the viewer.

A temporal redundancy occurs because the pictures in a video sequence tend to be similar (except in the case of a scene cut). Therefore it is often possible in an image sequence to predict the next picture from the preceding ones. Motion within the pictures must be taken into account. This makes the generation of a suitable prediction a computationally intensive process. Spatial redundancy in a picture occurs because pixel belong to regions with a similar appearance, for example a blue sky. Hence pixels can be predicted from surrounding pixels. In addition, the human eye is not very sensitive w.r.t. to the sharpness of edges, i.e., the contents of a picture is normally recognized even if the picture is a little bit blurred. Depending on the desired level of quality, the color representation, etc. can also be reduced, or a subsampled version of the picture can be transmitted. Thus a gracefully degraded version of the picture can be used which requires only a small data rate for the transmission but a human observer will hardly notice the difference from the original picture.

The principal structure of an H.261 or MPEG codec is shown in Fig. 2.2. A JPEG codec does not use a motion compensation. Therefore the JPEG coder consists only of the DCT, the quantization, and an entropy coding. The JPEG-decoder contains the respective inverse functions. The standards are similar in there principal structure, but differ in implementation details. They are not exact subsets of each other. The main tasks for exploiting the redundancies work as follows:

Temporal redundancy. This is the most efficient part for compressing motion pictures. JPEG does not use this compression technique because it was originally designed for still picture compression.

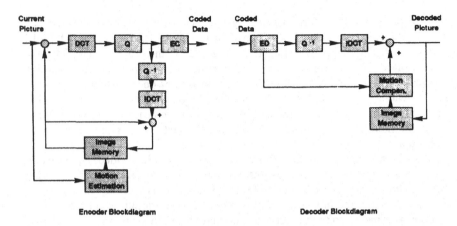

Figure 2.2. The principal structure of an H.261 or MPEG codec, where DCT (Discrete Cosine Transform), Q (Quantization), EC (Entropy Coding), Q-1 (inverse Quantization), IDCT (Inverse DCT)ED (Entropy Decoding).

The picture is subdivided into blocks of pixels (typically 16 x 16 pixel). The simplest prediction for such a block is its direct counterpart in the preceding picture. This prediction block is subtracted from the block of the current picture and the resulting prediction error is transmitted. If the pictures are very similar, i.e., there were no changes in the picture sequence (e.g., a fixed background), then the prediction would exactly match the current block and no data must be transmitted. However in real picture sequences the results would not be very good because certain parts of the picture have changed due to movements of the camera or movement of objects. This bad prediction can be avoided if the motion is estimated before generating the prediction. Typically, blockmatching algorithms [145] are used for this purpose. That is, a search area is defined in the previous picture. The block of the current picture is matched against all blocks within this search area. A search criterion is computed for each match (normally the $l1$ norm, i.e., the sum of the absolute pixel differences), and the best matching block, w.r.t. the search criterion is, used as the prediction. The encoder transmits again the prediction error and the displacement vector of the best match block. This full search blockmatching requires a high number of arithmetic operations. Array processors are well suited for implementing these algorithms [156]. In addition there are more sophisticated search strategies that save operations at the cost of a more irregular control flow.

Spatial redundancy. The spatial redundancy is exploited by a frequency transformation, the DCT (Discrete Cosine Transform) [167] in case of the abovementioned standards. The low frequency parts of the picture contain the main information. The high frequency parts contribute mainly to the sharpness of the edges. A low pass filtering of the picture gives a blurred version of the picture, whereas a high pass filtering yields just the contours of the picture objects. In this way the DCT can be used to split the image blocks into components that are important (the low frequency part) and into less important parts (the high frequency components). Furthermore, the pixels are de-correlated. As an example, a block of gray pixels with similar brightness would be translated into a single frequency component at the frequency zero. All other components have the value zero. Hence, a single none zero coefficient would be sufficient to replace a complete block of pixels in the transmission. The DCT is applied to blocks of 8 x 8 pixels. This is a compromise between the number of operations required for the transformation and the side effects (aliasing, noise) due to smaller windows. After applying the DCT only a few low frequency coefficients are necessary for transmitting the main image contents of the pixel block. For example, if the block corresponds to a uniform surface, only a single DC coefficient would be transmitted after the transformation, instead of the 64 coefficients which were necessary before applying the transformation.

Graceful degradation of the image quality. A quantization is used for this purpose, i.e., each coefficient is divided by the quantizer step size. This reduces the number of possible representations for a data coefficient. The human eye has only limited capability of distinguishing, e.g., certain gray scales. Furthermore, reducing the resolution of the high frequency components has only a small impact on the image quality. This is exploited in the JPEG and the MPEG standard (the quantizer step size is multiplied by a frequency dependent weight factor). Therefore the quantization saves data rate without sacrificing too much image quality. In addition, the quantization allows to control the data rate which is produced after the encoding: if the quantizer step size is increased, the data rate is reduced at the cost of a lower image quality. Thus the quantizer step size can be changed in a way to obtain an approximately constant data rate after the encoding. This is important to achieve a constant network bandwidth. Normally large changes in the consumed bandwidth are undesirable because they aggravate the problems of allocating transmission capacities to network users.

Entropy Coding. Standard source coding techniques like run-length coding and Huffman coding are used for a further reduction of the bandwidth. These techniques exploit the different occurrence probabilities of the data coefficient combinations. For example, large runs of zeros are very likely after the quantization. By assigning short codewords to these likely data words, and assigning long codewords to unlikely data coefficients, the average codeword length can be reduced. In this way the statistical similarities between image sequences are exploited to achieve a further reduction of the data rate.

Table 2.2.   DCT based standard compression algorithms.

Standard	References	Output Data Rate	Compression Ratio
JPEG	[154] [192] [95] [105]	–n. a.–	10 ... 20
H.261 or $p \times 64$	[125] [97] [96]	$64kBit/s$ ... $1.92Mbit/s$	10 ... 100
MPEG	[55] [94] [142]	MPEG-1 $1.5Mbit/s$, MPEG-2 $4 ... 10Mbit/s$	20 ... 100

The details of the different tasks are not considered in this thesis. It should be noted, however, that the standard algorithms differ from each other, although they use the same set of basic compression steps. Furthermore, there is a significant freedom within each standard concerning the implementation of each tasks. For example, there are many different block matching algorithms. Each of them can be used. The compression standard defines only the syntax for encoding the displacement vector and the type of motion compensation. Table 2.2 summarizes the main information concerning the different standard compression algorithms and their application spectrum.

Several public domain programs are available that implement the standards. Table 2.3 lists the programs that are used for the investigation described in this thesis. This already points out one main advantage in using C as the input to the design flow: since C is a widely accepted programming language, there is a large program base. Thus complex programs can be directly used for the system design. This reduces the development costs and increases the design productivity [196]. In practical designs, this more than compensates for the

Table 2.3.   Different public domain video compression programs
(available from ftp: //havefun .stanford .edu /pub).

Standard (Program)	Written by	Features	Lines of C code *)
JPEG (jpeg)	A. C. Hung, PVR-Group at Stanford University [85]	codec	8100
H.261 or $p \times 64$ (p64)	A. C. Hung, PVR-Group at Stanford University [86]	codec	6600
MPEG (mpeg_play)	K. Patel, Lawrence A. Rowe, Ketan Patel, Brian Smith, UC at Berkeley [149]	decoder	6000

*) The display functions are excluded from the line count.

disadvantages that C may have as seen from a theoretical specification point of view.

This section gave a brief description of standard algorithms. As pointed out before, these algorithms are not the only possible compression algorithms. Especially in the context of HDTV [147] or wireless communication [135], different algorithms may become popular that avoid some of the drawbacks of the DCT based algorithms. For example, DCT based algorithms tend to introduce block artifacts which are especially disturbing in case of high image qualities. Wavelet or subband coding algorithms avoid these artifacts. Another alternative are fractal based coding techniques. In comparison to JPEG, the fractal coding produces images which are more pleasant for a human viewer, in particular if very low bit rates are used [46]. Table 2.4. summarizes alternative implementations to the coding tasks specified in the standard compression algorithms. This list is by no means a complete list of alternatives. Rather it summarizes some of the most popular research topics in video processing (a good pointer to the main papers in video compression is [166]). It is important for the designer of the video compression system to know about these alternatives. Any of these alternatives might gain more popularity in the future. The video compression system should provide sufficient flexibility to support different alternatives, at least with a reduced performance. The next section outlines the basic structure of such a flexible compression system.

Table 2.4.   Alternative realizations of the compression tasks.

Compression Task	Possible Realization
Motion Estimation	Hierarchical blockmatching [48] [20], phase correlation [48]
Reduction of	Wavelets [6] [39] [42] [115] [129], subband coding [211]
Spatial Redundancy	LOT [101] [130] [214], fractal compression [16] [98] [99]
Entropy Coding	Vector quantization [57] [63] [36], arithmetic coding [207]

EMBEDDED VIDEO COMPRESSION SYSTEM

The video compression algorithms described above require high processing power and high memory bandwidth [156] [149]. Performing these tasks directly on the CPU of a PC will block the CPU for all other applications. This is not desirable because the user will perform others tasks in combination with the compression algorithms. For example in a video conference, the user might like to show and manipulate documents or visualize simulation results. The computer should handle the video conferencing and the document manipulation at the same time without stopping one application in favor of the other (or slowing them down significantly).

The problem can be solved if the video compression is performed by an embedded system. The system is integrated into the PC. It acts on behave of the CPU. The PC is the host system which provides the operating environment for the embedded system. The embedded system performs all the computationally intensive tasks. The local memory of the compression system is used for storing the intermediate results of the video compression. The video compression does not consume processing power or memory bandwidth from the PC. Hence the normal usage of the PC is not disturbed by handling the video data. The PC is used only for managing the video applications but all the video processing is performed by the embedded system. The partitioning of the tasks between the host system (the PC) and the embedded system (the compression system) is shown in Fig. 2.3.

The host system performs all tasks necessary to start and control the video applications. The PC will fetch the compressed video data from the network or the file system, it will select the appropriate video application (e.g., video

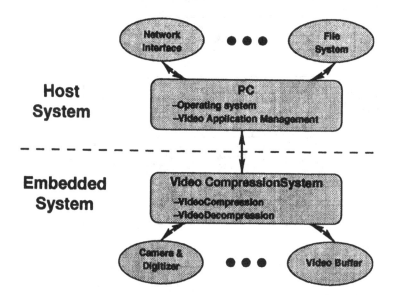

Figure 2.3.   Partitioning of tasks for the embedded video compression system and the host system.

conferencing, display, etc.), and it will manage the video display, i.e., it defines the position where to display the video data. However, it need not handle the uncompressed video data or process the compression algorithm. All this is done by the embedded video compression system. Both systems profit from the partitioning of tasks. The host system is not blocked by the video tasks. It can handle other tasks in parallel. The embedded system profits from the infrastructure which the host system provides. It needs no resources for file management, network interfacing, graphical user interfaces, etc. The embedded system can be tailored exactly to the compression tasks.

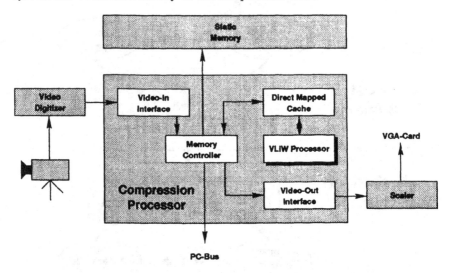

Figure 2.4.   Structure of the embedded video compression system.

The overall structure of the video compression system is shown in Fig. 2.4. The compression system consists of the compression processor (CP), the video digitizer that transforms the video data of the camera into a digital format, the video scaler that resizes the video data in such a way that they fit into the display window, and the static memory which is used for storing intermediate results (e.g., previous frames). The main components of the CP are:

VLIW processor. The VLIW (Very Long Instruction Word) processor performs the main compression and decompression tasks. The VLIW architecture has two major advantages. First of all, it supports a high degree of instruction-level parallelism. This allows to achieve high processing power for the video compression. The second major advantage is the possibility to parameterize the architecture. These parameters are used to select the basic structure of the processor (see section 2.4). This gives the opportunity to adapt the processor to different application spectrums. The development of the VLIW processor will be described in detail in following sections. However, the general design methodology is not restricted to VLIW processors. Other parameterizable architectures might be used as well. An example is a superscalar processor where the designer can select the number and type of functional units [103].

Data cache. The cache serves as a fast memory for the VLIW processor. Intermediate results of the compression are stored in this cache. A cache was chosen because it agrees with the memory model used by most compilers. This contrasts with the assumption in many videoprocessors (see section 1.1), where typically only a local memory is used. In that case the compiler would be responsible for allocating the memory blocks, i.e., the compiler has to manage the local memory like a register file. Normally this is not done by modern compilers. A special compiler must be developed. This reduces the design efficiency.

Memory controller. This unit performs the memory management tasks. It involves the managing of the different data streams in the video compression system. Data from the camera must be stored in the static memory, the decompressed video data must be transferred to the video buffer, access of the PC to the memory must be granted (e.g., to allow the PC the storing/retrieving of compressed video data in a DMA like fashion), and cache misses must be handled.

Video in/out interfaces. The interfaces contain FIFOs for buffering incoming and outgoing data. This gives the memory controller more freedom in handling the different data streams. The memory controller can poll the interfaces for new data at regular intervals. The memory controller must not react immediately upon new data coming from the camera. In addition the interfaces perform a subsampling of the video data. For example, in case of format conversion from the camera format to a smaller intermediate format, this is directly done in the interface. Thus, the required memory bandwidth to the memory of the video compression system is reduced.

The compression of video data works as follows. The data are recorded by the camera and transformed into a digital representation by the digitizer. The digital video data are given to the video-in interface. When a subsampling is used, the interface performs the necessary filtering of the data. The data are stored in a FIFO within the interface. The memory controller fetches a block of input data from the interface and stores it in the static memory. Currently it is assumed that an external chip performs the motion estimation (not shown in Fig. 2.4) or the motion estimation is switched off for the compression. This allows to realize the video compression system as a low-cost system. If a higher image quality and high compression is desired, an external motion compensation chip might be added. However, for all users that do not need such advanced features, the system costs can be kept low by using just the basic version of the video compression system. After the optional motion estimation is performed, the data are transferred to the VLIW processor that performs the remaining part of the compression algorithm. The compressed video data are transmitted to the PC. The PC can store these data in a file or it can transmit the data via the network. This will depend on the multimedia application in which the compression system is used.

The decompression of the video data works as follows. The CPU of the PC receives the data from the network or it fetches them from a file. The data are stored in the memory of the compression system. The memory controller loads these data from the memory into the data cache. The VLIW processor processes the decoding algorithm. The reconstructed picture is transmitted to the video-out interface. The interface will perform the interpolation of the video data. This is necessary, when a subsampled version of the image was used for the compression. The data are stored in a FIFO until the scaler fetches them. The scaler resizes the video data in such a way that they fit into the window assigned by the window manager. The data are stored in the video buffer at the appropriate position.

This section has outlined the main structure of the embedded video compression system. The details of the system structure are addressed in [198]. In particular, the parallel access to the static memory is described. This allows to transfer data and instructions to the compression system, and it provides a way to communicate with the host system. A more refined block diagram of the compression processor is given where the bus widths of the internal busses are specified. In addition, the instruction fetching of the VLIW processor is explained.

Up to this point, the design is mainly based on a qualitative analysis of the requirements in multimedia and on general assumptions about market developments for computer systems. The basic idea about the system to be designed is now specified in terms of a top-level block diagram. Typically this is the starting point for the real design, i.e., non-technical considerations have determined most of the first design decisions. Now each of the modules must be developed in a structured top-down process. The main component to of the design is the VLIW processor. The design automation for supporting an efficient design of this processor is the main focus of attention in this thesis. The next section defines the general structure of the VLIW architecture.

VLIW PROCESSOR ARCHITECTURE

Most of the processing in the compression system is done by the VLIW (Very Long Instruction Word) processor. This section presents the principal architecture and compares it to other related processor architectures. Furthermore, the main parameters of the processor which need to be defined during the design are summarized.

Two main forms of parallelism are used in current computer systems: data-level parallelism and instruction-level parallelism. In data-level parallelism, large arrays of data are processed in parallel on different processors. This is used for example in large scientific applications [51] like particle simulations, weather forecasts, etc. Most general-purpose applications, however, do not operate on such large data sets. For these application instruction-level parallelism is more appropriate. In this case several instructions are processed in parallel. Sometimes a different terminology is used for the different types of instruction-level parallelism, therefore, the definitions from [168] [103] [76] are recalled. These definitions will be used in the thesis (see Fig. 2.5):

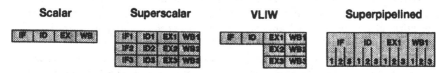

Figure 2.5.   Different types of instruction level parallelism.   IF (Instruction Fetch), ID (Instruction Decode), EX (EXecute), WB (Write Back).

- A **scalar** processor fetches one instruction per cycle, but overlaps the processing of several instructions within a RISC-like pipeline [76].

- A **superscalar** processor fetches several instructions from a scalar instruction stream and processes these instructions in parallel. The hardware of the processor checks the dependencies among the operations, reorders the instructions if necessary, performs register renaming, and executes them in parallel if possible [103] [43]. No specific compiler support is required in this case. In fact, a binary code compiled for a scalar processor can be used directly by a superscalar processor with the same instruction-set.

- A **VLIW** processor fetches one long instruction word per cycle [168]. This instruction word consists of several operations that are executed in parallel. The compiler is responsible for all dependency checks and for reordering the operations. In comparison to a superscalar processor the instruction fetch and decode is simplified at the cost of a much more complex compiler support.

- A **superpipelined** processor uses several intermediate pipeline stages for each stage of a traditional RISC-like pipeline [76] [139]. This allows to increase the clock frequency but it also requires a faster memory and it is more susceptible to pipeline hazards.

The main advantage of the VLIW processor is the simple hardware structure especially in the instruction decoder. This allows to make the VLIW processor parameterizable as described below. Fast processors can be designed. The structure of the processor is very clean. The main disadvantage is the necessary compiler support [195]. VLIW processors rely heavily on the compiler to achieve high performance. Only static parallelism can be exploited, i.e., all dependencies must be resolved at compile-time. A conservative decision must be used in critical cases. This can reduce the potential parallelism. Different generations of VLIW processors are not binary compatible. This is not an important issue for embedded system design since the code for embedded system will normally be recompiled anyway, but binary compatibility is an important topic for general-purpose processors [195].

The generic structure of a VLIW processor and a typical instruction word are shown in Fig. 2.6. The processor can be subdivided into two main parts: the control unit and the datapath. The control unit fetches one instruction per cycle. One part of this long instruction word controls the fetching of the next instruction. This part includes the jump condition (*Jump_cond*) which determines the flag that is used as the condition of a branch, and the jump address (*Jump_addr*) that determines the address displacement of a taken branch. When the branch is not taken, the next instruction is fetched. The remaining part of the instruction word controls the datapath. Each unit of the datapath has its own field in the instruction, i.e., the instruction is completely orthogonal. Instructions from different units do not interfere with each other. The datapath is directly controlled by the instruction word. No further micro-program look up is required. The number of units in the datapath can be increased. For each new unit, a field is added to the instruction word. This makes the VLIW architecture well suited as a generic architecture template.

The following parameters must be specified for the two components of this template. The control unit is defined mainly by the width of the control word, the number of flags, and the number of words in the instruction RAM or in-

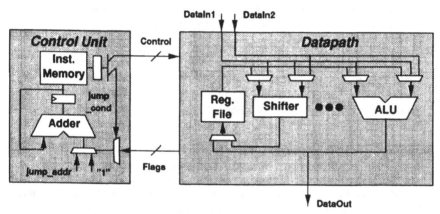

Instruction word:

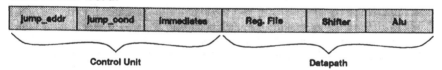

Figure 2.6.   Structure of a VLIW processor and typical instruction word.

struction cache. The datapath is the main part to be configured, the following parameters must be selected in the design process:

Number and type of the functional units. Each of the functional units performs one or more RISC-like operations, for example, addition, multiplication, shift, etc. The units are assumed to have a delay of one clock cycle. A new operation can be started on (or *issued* to) a unit in each cycle. The units might be pipelined. The degree of pipelining, i.e., the number of pipeline stages in the unit, defines the latency of the unit. That is, the time in clock cycles after which a result can be accessed at the output of the unit. The latency is a variable in the architecture template.

Registers. The size and the number of read/write ports in the register file can be specified. Currently, only a single general-purpose register file can be used. This restriction is due to the MOVE scheduler [34] [82] [81] which is used in the compiler back-end. In the future this restriction might be removed so that several small register files can replace the large register file. This would be more efficient in the synthesis of the processor hardware.

Number and type of ports. The datapath has different types of ports. The input and output ports are used for loading and storing data into the memory. These ports consist of two physical ports. One port is used for the data transfer and the second port is the address port which specifies the memory location. The immediate ports transfer constants from the control unit to the datapath. The immediates are part of the instruction word. Hence, they do not require a special address. Finally, there are ports for reading and writing the program counter and for accessing the return address register. These ports are used in case of a function call to jump to the first operation in the function and to return to the point where the function was called.

Interconnection network. Typically a restricted interconnection network will be used in the datapath, i.e., not all units are connected with each other. The restricted connection saves area on the chips real estate and, more important, the time for routing the data is reduced, and hence the processors cycle time is improved. A single driver is assumed for each connection, i.e., no tri-state busses with multiple sources are used. This simplifies the hardware synthesis but some overhead in terms of area might occur due to the larger number of busses.

This target architecture will be used for the design flow described in the next section. Such an architecture template is necessary to implement a design flow, but it is just one example of a possible processor architecture. In fact most modern microprocessors are defined in terms of a generic architecture with specific implementations. Examples include x86 [76], PowerPC [44] [45] [141], Alpha [177], PA [120], MIPS [109], SPARC [186], etc. These architectures can also be used as building blocks for the design flow. In principal, they are variations of Johnson' s generic superscalar architecture [103]. They differ mainly in the names of the instructions and in the specific way of handling interrupts. Comparing the architectures to the VLIW template shows many similarities. The main difference is the instruction fetch. Superscalar processors require some form of reorder buffer and reservation stations. Synthesizing these components directly with a normal ASIC library would be too costly. These demanding components must be available as pre-defined full-custom elements [9]. In that case it would be possible to include superscalar processors into the design templates. This section will use only basic hardware synthesis. Therefore the VLIW architecture is used, but the design flow is kept general so that other templates can be included in the future.

3 DESIGN METHODOLOGY

The previous section gave a summary of the relevant background material. It addressed the real-time condition which is an important design constraint. The performance, which is at least required from the embedded system, is defined in this way. The application range of the embedded system was investigated and the major compression standards where reviewed. The principal system structure was motivated: a flexible, processor based compression system which is integrated into the video card of a PC. The core of the system is a VLIW processor.

CHALLENGES OF SYSTEM-LEVEL CODESIGN

This list of different topics is already an indicator of the complexity that is involved in the design of embedded systems. The complexity stems from the fact that many disciplines have to be considered in a system-level approach. Each part allows many different variants and modifications. The main design task is the search for the best combination of the variants. On lower levels of abstraction the design often involves only the optimization of certain predefined structures. On the system-level it is in most cases not possible to formulate all the different influences in form of a mathematical cost function. Rather, it is necessary to include much more knowledge into the design process. This knowledge is the crucial ingredient for making a good design. The designer must know about all the alternatives and he/she must have a way to evaluate their possible trade-offs. The measuring of the trade-offs is done by the tools. This means the tasks of the design tools shift from low-level optimizations into providing information about trade-offs for many different system variants. The tools have to extract data and they must prepare them for the designer.

This section takes a look at the new levels of complexity that are encountered in embedded system design. Next, the general design approach is motivated. It features the combination of the tool's ability to gather information with the designer's knowledge about the design context. Finally, the detailed steps of the design flow of the video compression system are presented. The complexity of the design manifests itself at three main points:

- A larger number of operations is mapped on a functional unit. The input specification does not reflect the structure of the system any more.

- A higher degree of flexibility is required. Therefore the software development is a significant part of the design. The higher flexibility allows to address a complete range of applications in contrast to just a single application.

- A combination of optimizations from hardware, software, and algorithms is the main design goal. In contrast, designing on lower levels of abstraction focuses on these domains individually. But the combination of optimal parts will not necessarily lead to a good system. A more global perspective is required.

The next paragraphs look at these points in more detail. The first point is the different complexity, e.g., of system-level design and high-level synthesis. High-level synthesis first constructs a CDFG (Control Data Flow Graph) from a given specification. Next, a suitable datapath is derived by scheduling and allocation [169]. The datapath closely reflects the structure of the input specification. It is more or less directly "compiled" into silicon. This is possible, because only a small number of operations is mapped on the hardware. For example, the elliptic wave filter, which is frequently used as an example in high-level synthesis (e.g., [74] [89] [114] [151] [213]), has approximately 34 operations. These operations are mapped onto a datapath of more than 2 units. This means, less than 20 operations are mapped onto a unit. Therefore the hardware structure can be adapted directly to the communication pattern of the CDFG. The designer has much control on the final hardware when writing the initial specification. The situation is different for the video compression programs. A much more abstract approach is necessary since the complexity of the compression programs is several orders of magnitude higher. For example the *mpeg_play* program consists of approximately 20k operations which are mapped onto approximately 10 units. In this case more than 1000 operations are mapped onto a single unit on average. It does not make sense to examine each of these operations directly when deriving the datapath. Instead, a more abstract way of representing the program requirements is necessary. One such way is a quantitative analysis [76] that will be used below. It was shown that this is a very powerful concept for designing general-purpose processor. Applying it to embedded system design means to create design tools for measuring and visualizing the analysis data [201] (see chapter 4) and chapter 5.

The second problem is the software development. As mentioned in chapter 2, the video processors are typically very complex. They use a high degree of instruction level parallelism. The video application are also very large. This means, complex programs for complex processors must be developed. This is normally a tedious and time-consuming task. For example, in [202] [197] approximately 10% of the time was spend for developing the concept, 30% of the time was spend on creating an RT level model of the hardware, and 60% of the time was spend for developing the assembler code for some important functions of the video compression algorithms. Similar findings were obtained in industrial surveys [152]. This high amount of development time for the software is due to the complexity of the processors and due to the high simulation times. Typically the RT level model of the processor must be used for testing and debugging the opcode. A model written on an higher level of abstraction does not reveal the bugs due to the pipelining and due to the complex internal structure of the processors. But using the RT level model results in extremely high simulation times which make debugging and testing of a complete video compression application nearly impossible [1]. For example, about 1 cycle/s can be simulated for a VHDL or Verilog description of a video processor on a SPARC IPC like workstation. Simulating a short MPEG sequence (e.g., the *flower.mpg*) sequence will require in the order of 10^8 processor cycles. Hence simulating the *flower.mpg* sequence would require 3 years. The codesign approach, outlined in the following, offers an alternative because most of the programming is done in C/C++. The normal programming environment can be used for debugging and testing the programs.

The code can be ported to many different variants of the processor structure. This is already a crucial step to achieve the key feature of the system-level design: the combined design of hardware, software, and algorithms. It is a result of the more abstract design perspective. The designer can produce the variants more easily. This allows to investigate the trade-offs of many different variations. A conventional design would not analyze them in detail due to a lack of design time. While this abstract approach gives more freedom to improve the design quality by checking various solutions, it also requires much more interdisciplinary knowledge. This must be integrated into the design flow. The result is a designer centered approach.

DESIGNER CENTERED APPROACH

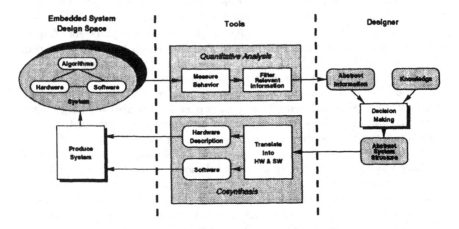

Figure 3.1. A designer centered approach. The tools are used to analyze possible system variants. Next, the designer decides on the appropriate system structure which the tools translate into the final system specification.

The designer centered approach combines the designer's knowledge with automation by design tools (Fig. 3.1). The tools are used at two key points of the design. First, they extract the information about the design space. This is done by the *quantitative analysis*. Each step of the analysis is divided into two phases: (1) the measurement of the data and (2) the preparation of abstract views. The abstract views concentrate the information for the designer. He/she takes these data and combines them with his/her knowledge about the design context. This is the basis for making the main design decisions. The result is a coarse definition of the system structure. The structure is only specified on an abstract level. No implementation details are worked out by the designer.

This is the second point where the design tools increase the efficiency of the design process. The *cosynthesis* is used to translate the abstract specification into the detailed specification. This includes both, the hardware description and the software. The combination of both parts is the foundation for producing the final system. They can form the basis for developing the next generation of the system. This is another advantage of designing on a more abstract level. It is easier to port the design specifications since they are less implementation dependent. For example, VHDL hardware blocks and C software programs can be used for many different designs. This is less specific than assembler programs or dedicated hardware.

DESIGN FLOW

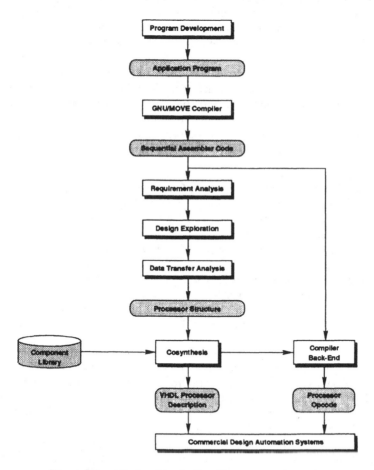

Figure 3.2.   Design flow for the video compression system.

The general approach can be refined into specific design steps for a certain application domain. Fig. 3.2 shows the overall design flow for the compression system [200]. The design process starts with the development or reuse of compression programs. The programs are translated into sequential assembler code. This assembler code is based on the final instruction-set. But timing information (pipelining of units or branch delays) is not considered. The assembler code is the main input for the analysis. It allows to investigate the programs in detail, but it does not require all the implementation details of the hardware. The quantitative analysis is broken down in three parts: the requirement analysis, the design exploration, and the data transfer analysis. Each of these steps reveals more and more details of the best suited hardware structure. The result of the analysis is the coarse definition of the processor structure. It is the main input of the cosynthesis. This step transforms the processor "skeleton" into a VHDL description and it creates a machine description for the compiler back-end. In conjunction, the parts define the hardware and the software of the system. They are comparable to manually written specifications. It is possible, to use them directly as input for commercial design tools. The hardware can be synthesized and cosimulation can be performed to validate the design. In more detail, the steps look like this:

Development and adaptation of C/C++ programs. The programs can be developed with the conventional programming and debugging environments. The functional testing of the programs is done by compiling them for a workstation or PC, which is used as development platform. The programs are executed like any other program. This is a very fast way of "simulating" the video applications, e.g., the *mpeg_play* program achieves approximately half the real-time speed on a SPARCstation10 (CIF format, 352x288 pixel, was used for the video sequence). This even allows to view the video sequences. The image quality can be directly inspected by the designer. It facilitates the evaluation of algorithmic changes on the image quality.

Furthermore, existing programs can be used. The analysis presented in chapter 7 is mainly based on public domain C programs. In this case it is not necessary to develop a complete program from the scratch. Only minor modifications must be performed to adapt the programs to the structure of the planned embedded system. In case of the *mpeg_play* program, the dithering and the X window display were switched off because these tasks are handled directly by the scaler of the video compression system. This means, design time is focused on improving the critical parts of existing programs rather than "re-inventing" them for a slightly different processor architecture.

Translation of the program into a sequential assembler code. The GNU / MOVE compiler [82] [81], developed by J. Hoggerbrugge at Delft Technical University, is used in this thesis. The compiler is based on the GNU gcc/g++ with a special back-end to generate MOVE assembler code. For the purpose of this thesis the MOVE code can be considered as a generic 3-operand RISC code with no special addressing modes. The specific features of the move architecture [35] [183] [182] (e.g., transport triggering, hybrid pipelines, etc.) are not used. Other compilers can be used if a simulator exists for profiling the opcode. Examples are the MIPS compiler with pixie [180], SUIF [205] with pixie, etc.

Requirement Analysis [76] [201]. As emphasized above, the compression programs are complex (at least when considering them as input specifications for design flows). Compiling them into assembler code includes many transformation in the compiler. This makes it next to impossible to predict the performance by inspecting the algorithms. The interactions between the compiler, the specific software implementation, and the algorithms are too complex. Therefore a precise measurement is necessary. The programs are simulated with representative input sequences. The execution frequencies of the basic blocks are recorded. These data allow to calculate instruction mixes, to determine the most demanding functions, etc., as described in chapter 4.

The data are a basis for improving the algorithms and their software implementation. Often, a redesign of the most crucial parts is sufficient. In addition, lower bounds on the required number of operations (they exclude delays due to data dependencies or resource conflicts) are calculated. This in turn defines the requirements for the minimum number of functional units that must be used in the processor. The bounds are useful for restricting the design space of the next analysis step.

Design exploration [27] [199]. This step measures the achievable performance w.r.t. the chosen processor architecture and the properties of the compiler back-end. All parameters of the architecture template can have an influence on the performance. Each of these parameters is a degree of freedom in the processor design space. A complete exploration is normally prohibitive due to the resulting processing-times. But the design space is highly non-linear, i.e., all parameters interfere with each other, and the resulting performance can not be determined by a simple interpolation from a small number of known design samples. A systematic approach is mandatory [165] [116] [104] [100] [175] [176]. Each step of the exploration generates bounds. The bounds permit to reduce the design space while focusing on its essential parts (see section 4.3). The bounds are established via relaxed processors. These relaxed processors exclude certain constraints on the hardware. For example, the interconnection network is excluded during the design exploration. A complete interconnection network is assumed. The register file size might be unrestricted, etc. Gradually, these relaxations are removed, as more and more parts of the processor structure are defined by the design process. The unlimited number of units will be replaced by a specific unit constellation, the unlimited register file size will be set to a reasonable value, etc. This will typically reduce the performance. The scheduler has to consider more constraints, which reduces the freedom for parallelizing operations.

The designer can use the data of the relaxed processors as bounds. None of the more specific processors will achieve better performance than a relaxed processor. Therefore, a guideline for the optimization is given by gradually refining the relaxed processors. The reduction of the design space is controllable without implementing all possible processor variants. The designer can focus on specific processor structures without neglecting the evaluation of possible design alternatives.

Data transfer analysis. Typically the interconnection network of a datapath has a significant influence on the performance. The performance will severely degrade if, for example, frequently used data transfers are not supported by direct connections among the functional units. On the other hand, an oversized interconnection network occupies too much chip area and would cause a large delay which slows down the processors.

Therefore it is important to analyze the data transfers between the operations. The most frequently used transfers will be supported by direct connections between the units, whereas seldom used transfers are handled by storing the intermediate results in the register file. The designer can trade-off between the utilization of the functional units and an increase of the network size.

Cosynthesis. The results of the previous analysis step defines the principal structure of the processor. This processor structure is entered on a block diagram level into the CASTLE system. A schematic entry facilitates the process. The CASTLE cosynthesis environment generates a synthesizable VHDL description of the processor and a corresponding compiler back-end from the schematic. The hardware and the software is generated from a single description. Thus the design tool ensures the consistency between the processor hardware and the compiler. All this is done automatically without any manual design work. Several processors can be developed. This allows to study trade-offs between processor performance and complexity. Such a detailed investigation is very important, because often small changes in the processor architecture can improve the performance/cost ratio significantly.

Final hardware implementation. The best processor from the cosynthesis step is selected for the final hardware implementation. This step is similar to a conventional processor design from an RT level hardware description. Demanding components like register files can be realized by full custom designs while all the remaining components are synthesized by commercial design tools.

Cosimulation. This step validates the synthesized processor in conjunction with the generated opcode. A commercial VHDL simulator is used which simulates the processor while executing the opcode of the application program. The results of the simulated program can be compared against the results obtained from the simulation on higher level of abstraction. For example, a block of pixels which is calculated in the cosimulation is checked against a block calculated during the execution of the initial C program. In this way the correctness of the processor and the opcode is checked for a number of different data sequences. The step is extremely time consuming for the video applications but most of the processing can be done automatically, without interference by the designer.

The proposed codesign approach aims at a systematic refinement of the design space spanned by hardware, software, and algorithms. The analysis steps should not be seen independently from each other, since the design space is non-linear. The most suitable design decisions in one dimension depend on the decisions in another dimension. For example, the most suitable algorithm depends not only on the achieved image quality but also on the chosen hardware structure [135] [41]. A high image quality is not very interesting if it can only be achieved with unacceptable system costs. This means all the design decisions must be seen in the light of the specific embedded system design. There is no single best algorithm or processor structure. Rather, the most suitable combination w.r.t. the current design constraints is the most important thing. Therefore a number of possible alternatives will be explored at each step. From these alternatives a subset of the best implementations is selected for the next design step, resulting in a stepwise refinement of the design space. In addition a frequent back tracking and reconsidering previous design decision based on the results of the current design step will be necessary (a framework which supports this linking of analysis results is presented in chapter 6). This means, a lot of information about the different design variants is created. The designer must find the important data. The tools will help him/her in evaluating all the information. This will be detailed in the following chapters. The next chapter looks at the information that is provided by the quantitative analysis.

4 QUANTITATIVE ANALYSIS

The main task of the system design is a careful balancing of all system resources to achieve a high performance / cost ratio. The quantitative analysis is the tool for measuring the performance before developing the system in detail. It refines the design space in a stepwise procedure (see Fig. 4.1). Each step provides more accurate bounds on the performance of the final system. It selects the most promising solutions for the next analysis step. The initial design space of the embedded system design is mainly defined by marketing conditions and existing products. Typically, a company will not start a product from scratch, neither will it work for a completely new market segment. Instead, the market segment defines the basic constraints on the system cost, e.g., PC component prices must be below \$1000. Furthermore, the system must achieve a minimum performance level, e.g., decoding of CIF sized images in real-time for frame rates of $30\,fps$ (frames per second). If the system cannot meet this basic goal, it cannot compete with other similar products. In summary, the principal system structure, the maximum costs, and the necessary performance are already defined. This gives the basis for the more detailed analysis, described in the following.

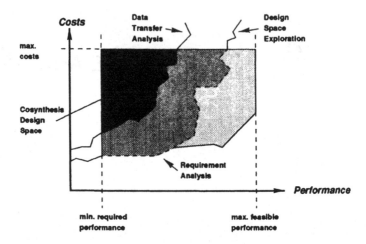

Figure 4.1.   Bounding costs and achievable performance by a stepwise reduction of the design space.

The first step is the *requirement analysis*. It mainly investigates the algorithmic properties and the implementation of the software. It neglects all performance inefficiencies that are due to the hardware implementation or caused by scheduling inefficiencies. Thus it determines the minimal processing power which is required for the system. A peak performance below this rate will not be sufficient to execute the algorithm in real-time. This minimal performance determines the minimal size of the processor and thus lower bounds the costs of the embedded system. The next step is the *design exploration*. It determines an upper bound on the performance that can be achieved by the processor architecture. This step considers the idle times of the processor units. Idle times occur due to resource conflicts or due to data dependencies. This is done by the compiler back-end. It reads the processor description for a specific processor structure. A scheduling and allocation is done. It assigns the operations of the sequential assembler code to certain time slots and certain units of the processor. The data dependencies are considered when assigning the time slots (pipelining of operations is taken into account). The resource conflicts are taken into account when assigning the operations to the functional units. The result is a parallelized program. The scheduled program is used to determine the processing time for different data sequences. The process is repeated for different processor structures.

Possible changes of the processor structure include different numbers of functional units, different register file structures, and different numbers of load / store ports (see section 2.4). A complete interconnection network among the functional units is assumed. The result of the design exploration is a set of processing times for a large number of different processor variants. This set is systematically reduced until only a few suitable processors structures are left. They are the basis for the investigations in the *data transfer analysis*. It investigates the interconnection network for these remaining processor variants. Again a reduction of the design space is achieved. Typically, one or two processor structures with roughly specified interconnection network will remain as a result of the data transfer analysis. These processors are implemented in detail by the *cosynthesis*. Again, the designer can check alternative solutions. This allows to investigate implementation details that were not evaluated by the more abstract analysis steps. The design tools help the designer because they facilitate the fast implementation of alternative designs.

The starting point for the quantitative analysis is the real-time condition defined in section 2

$$\forall \omega \in \Omega - S_{except} \; : \quad \frac{1}{blockRate} \quad \geq \quad \frac{time_\omega}{block}$$

$$\iff \; \forall \omega \in \Omega - S_{except} \; : \quad \frac{1}{blockRate} \quad \geq \quad \frac{cycles_\omega}{block} \cdot \frac{1}{f_{Clk}} \quad . \qquad (4.1)$$

The real-time condition states that the time between two data blocks $\frac{1}{blockRate}$ should be larger than the time needed to process a block $\frac{time}{block}$. The left side is defined by the image formats (see Table 2.1) and the right side depends on the implementation of the video compression system. The inequality must be correct for all but the permitted exception sequences, i.e., it is fulfilled with probability $1 - P_{except}$. The sequence index will be dropped in the following, if the specific sequence is not important. The right side of the inequality can be further refined by splitting it in the number of cycles per block ($\frac{cycles}{block}$) times the execution time for each cycle ($\frac{1}{f_{Clk}}$). Specifying these values would require a complete system implementation. The $\frac{cycle}{block}$ are determined by the software implementation and the compiler, while the f_{Clk} depends on the processor hardware. Designing a large number of systems just to check the performance is not feasible. Therefore the bounds on the performance are essential. This avoids the implementation of a specific system while providing data for refining the design space.

The first step is to look at the number of operations required by the algorithm. This is achieved by inspection of the sequential instruction stream produced by the compiler. The average pathlength for a data block is denoted by $\frac{op}{block}$. Furthermore, we can introduce the number of operations per cycle

OPC provided by the hardware. This defines the peak performance of the processor, i.e., all functional unit are used to perform an operation. However, without a detailed design we do not know the portion of this performance that will actually be used. This is accounted by a utilization $0 < util \leq 1$. It is unknown for the time being. The utilization defines the fraction of the peak performance that is used to perform useful operations. All other operations are essentially *nops* which occur due to data dependencies or resource conflicts. All the idle times are summarized in the utilization which permits to restate equation 4.1 as an equality

$$\frac{1}{blockRate} = \frac{op}{block} \cdot \frac{1}{util \cdot OPC} \cdot \frac{1}{f_{Clk}} \quad .$$

The main point of interest for the processor design is the most suited number of OPC that the processor hardware should provide. Thus we solve for the OPC to obtain the basic design equation. It will be refined during the quantitative analysis

$$OPC = \frac{1}{util} \cdot \frac{op}{block} \cdot \frac{blockRate}{f_{Clk}} \quad . \tag{4.2}$$

The requirement analysis establishes a first performance bound by assuming a utilization of 1. The right-hand side contains two unknown variables, $util$ and f_{Clk}. The $\frac{op}{block}$ is known from the simulation of the assembler code with different input sequences. The $blockRate$ is known from the real-time condition and the performance that is required from the system due to the market conditions. Assuming a utilization of 1 defines the minimum peak performance of the processor that is required to execute the video processing applications in real-time. The design exploration will be concerned with deriving more accurate bounds on utilization. The data transfer analysis extends these data by considering the probability of data transfers between the functional units and by measuring the number of these transfers that occur for a particular processor structure. The details of these analysis steps will be described in the following sections. As a whole they form a picture of the complete system design space (see Fig. 2.3). Even when using automatic optimization tools, it is important for the designer to know these data. Typically, the automatic tools can do some basic work, but the most significant improvements are gained from combining the results with the designer's knowledge about the application domain. This kind of knowledge is not available to automatic tools. The situation will be illustrated by looking at the IDCT (Inverse Discrete Cosine Transform). It is one of the most important functions in all of the compression programs. It will serve as an example to show the limitations of an automatic approach. It motivates the necessity of including the designer's knowledge into the design process. It shows why an analysis is so important. Only the analysis gives the designer an indicator of what happens in a complex design space. Hence, it is an essential tool for performing knowledge based decisions.

A matrix implementation of the IDCT written in C is discussed (Fig. 4.2). The example uses the full optimization capabilities of the GNU/MOVE *gcc* compiler. But even in this simple case it is possible to achieve an eightfold reduction of the number of store operations by a simple manual modification of the source code. In this case, no valid C compiler can perform the optimization, but it is easy for the human designer due to his/her knowledge about the application context. Two implementations of a matrix IDCT realization are compared *dctSimple* and *dctDis*.

```
void dctSimple(int* data, int* intermediate) {
static int idct_matrix[] = {
...}; /* IDCT-matrix multiplied by 2^8 */
int i, j, k;
for(i=0; i<DCTSIZE; i++)
  for(j=0; j<DCTSIZE; j++) {
    intermediate[j*DCTSIZE+i] = 0;
    for (k = 0; k < DCTSIZE; k++)
      intermediate[j*DCTSIZE+i] +=
        idct_matrix[i*DCTSIZE+k]
        * (int) data[k*DCTSIZE+j];
  }
  ...
}
```

```
void dctDis(int* data, int* intermediate) {
static int idct_matrix[] = {
...}; /* IDCT-matrix multiplied by 2^8 */
int i, j, k, temp;
for(i=0; i<DCTSIZE; i++)
  for(j=0; j<DCTSIZE; j++) {
    temp = 0;
    for (k = 0; k < DCTSIZE; k++)
      temp += idct_matrix[i*DCTSIZE+k]
        * (int) data[k*DCTSIZE+j];
    intermediate[j*DCTSIZE+i] = temp;
  }
  ...
}
```

Instruction Mix of IDCT Matrix Realizations

Inst Mix	Operations / 8 x 8 pixel									
	ADD	SUB	MULT	SHIFT	COMP	LOAD	STORE	LOGIC	FP	EXT	REST
dctSimple	7440	0	1024	2510	1170	3140	1220	0	0	0	0
dctDis	7190	0	1024	2320	1170	2050	128	0	0	0	0

Figure 4.2.   Realization of the IDCT by simple matrix multiplication.
DCTSIZE is 8, only the first matrix multiplication of the two-dimensional transform is depicted. Inserting a temporary variable temp in the *dctDis* function leads to a reduction of the store operations by a factor of eight.

Both implementation use the following basic equation

$$\mathbf{IDCT(A)} = \mathbf{T} \cdot \mathbf{A} \cdot \mathbf{T}^T ,$$

where **A** is an $N \times N$ block to be transformed. N is equal to 8 for the standard video compression algorithms. T is the kernel of the 1-dimensional transformation. Since the DCT/IDCT is a separable transformation, it can be performed in two steps, first in y-direction (first matrix multiplication by T) and afterwards in x-direction (second matrix multiplication by T^T). A straightforward C code realization consists of three nested loops for each matrix multiplication.

This is depicted in *dctSimple* of Fig. 4.2. The number of multiplications is given by $2 \cdot N^3$ which is equal to 1024. An output value is created after processing the innermost loop, i.e., after forming the dot product of two matrix vectors. Hence, the number of store operations should be one $8th$ of the number of multiplications, i.e., 128. The intermediate value of the accumulation should be kept in the register. The real value, shown in the table of Fig. 4.2, is nearly 8 times as high. An inspection of the assembler code for the inner loop body shows that the intermediate value is saved in memory after every iteration of the inner loop. This is necessary because the arrays are passed as pointers in C. In principal, the arrays might overlap in memory. This would be perfectly valid C code, although it is a bad coding style. The compiler cannot decide that the arrays are completely disjunct in this particular case. The designer can make this decision without any difficulties. He/she knows that a normal matrix multiplication was intended. Therefore no aliasing of the arrays will occur. The designer can express this kind of memory disambiguation by inserting a temporary variable. In this case the compiler can keep the temporary variable in the register file to avoid the unnecessary store operations. This is implemented by the *dctDis* realization. The number of store operations is now 128, as expected. It indicates the limitations of an automatic approach. The compiler *cannot* perform the transformation from *dctSimple* to *dctDis* because it is *not* valid for all C programs. The designer has more knowledge about the intended usage of the functions. This allows him/her to perform transformations which are beyond the scope of an automatic tool. The example proves the necessity of giving the designer a good picture of the design space. It is the main contribution of the analysis steps described in the sequel.

REQUIREMENT ANALYSIS

Example data of a requirement analysis are shown in Table 4.1. The operation count data are obtained by an instruction set simulation of the *mpeg_play* program [149] using the *moglie.mpg* sequence as data input. The instruction set simulator of the MOVE system was used. The table shows some interesting properties. First of all, the operations count is not equally distributed among the functions, rather, it is local to a few functions. The designer can concentrate on these important functions. The first three functions in the example consume approximately 96% percent of the total operation count. A second important information concerns the different types of operations that are used. This is called the *instruction mix*. It can be used to adapt the number of functional units to the required number of operations. A third property is the change of the instruction mix when comparing different functions. The IDCT function, *j_rev_dct*, requires many additions, many compares, and many load/store operations. The *ReconIMBlock* function, implementing the reconstruction, consumes mainly additions and load operations. It fetches the prediction block from the memory and adds it to the block that is currently decoded. The similarity or disagreement in the instruction mix is an important characteristic. It governs the performance that can be achieved when processing a number of different functions on the same hardware. Combining very different functions will lead to idle times. The instruction mix allows to quantify the mismatch effect.

Table 4.1.   Most important functions and instruction mix of the *mpeg_play* program when simulating with the *moglie.mpg* sequence as data input.

mpeg_play: Important functions

Functions	% of Ops	Ops/Macroblock	Instruction Mix						
			ALU	MULT	REST	COMP	SHIFT	LOAD	STORE
j_rev_dct	39.22	5046.8	41.51	6.63	9.64	11.97	9.91	11.39	8.94
ReconIMBlock	32.28	4154.0	51.37	0.14	0.29	10.21	0.43	28.31	9.24
ParseReconBlock	25.04	3221.7	40.31	2.21	0	12.86	14.88	17.00	12.75
ParseMacroBlock	1.68	215.6	35.83	0.46	0	21.79	8.98	24.50	8.44
j_rev_dct_sparse	0.48	62.1	45.83	0.00	2.78	2.78	2.78	1.39	44.44
bcopy	0.40	51.9	50.9	0.00	0	17.14	0.94	15.51	15.51
CollectStats	0.32	41.5	50.74	0.06	0.03	1.93	0.00	31.51	15.74
next_bits	0.18	22.7	42.9	0.00	0	14.20	17.46	25.45	0.00
init_stat_struct	0.12	16.0	52.5	0.00	0	5.16	0.00	0.08	42.26
mpegVidRsrc	0.04	4.8	24.47	0.00	0	66.66	0.66	5.82	2.38
vfprintf	0.04	4.7	39.84	0.00	6.26	26.39	3.74	13.11	10.66

In general, the requirement analysis shows the effect of mapping different applications on a fixed data path and it provides a way to estimate the effects of changing the algorithms or the software implementations. This section derives the main parts of the requirement analysis by a stepwise refinement of equation 4.2. The results of applying these tools to the analysis of the video compression system are presented in section 7. The most global result of the requirement analysis is the total operation count per block $\frac{op_w}{block}$, i.e., the pathlength for processing one data block using a specific instruction-set. This result permits to compare the overall complexity of different compression programs. It gives a first estimate on the necessary peak performance of the processor and it provides a way to select typical image sequences for the analysis. In theory, all possible image sequences must be analyzed to ensure that the real-time condition is fulfilled with probability $1 - P_{except}$. In practice this would be prohibitive due to the excessive amount of processing and design time that would be required for such an analysis. Therefore a practical approach will start with a large set of sequences, for example, the standard sequences which are used for analyzing video compression algorithms. These sequences reflect typical tests for different compression situation, e.g., constant motion versus fast changes, many image details versus large uniform surfaces, etc. The $\frac{op_w}{block}$

shows how complex it is to process each of the sequences. The distribution of the operation count allows to select a representative ensemble from the set of all image sequences. The ensemble will be the basis for all further analysis. The ensemble can be further reduced after gaining more insight into the performance of the embedded system.

Important Functions

As shown by the IDCT example at the beginning of this chapter, an important source for improving the $\frac{op}{block}$ is the manual optimization of the software implementation. It exploits the designer's knowledge about the application context. However, applying these optimizations indiscriminately throughout the program would be an immense waste of design time and thus increase the system costs. This can be seen by looking at the results of Table 4.1. In this case the dynamic operation count is very high for the first three functions. There is a high likelihood that an operation, which is fetched during the program execution, stems from these important functions. Optimizing them is a crucial point for a good program performance. But the likelihood of the other 150 functions is low, so optimizing of these functions would just increase the design costs without improving the overall performance. This means, the important functions must be determined before performing optimizations. Let F denote the set of all function in the program. The probability space of section 2 is extended by the function dimension, i.e., $(\Omega \times F, \mathbf{B}, \mu)$. $\mathbf{B}$ denotes the σ-field of subsets from the extended space. The probability of being in function $func \in F$, when processing the data sequence $\omega \in \Omega$, is denoted by $\mu(func|\omega)$. The operation count of the function $func$ is denoted by

$$\frac{op_\omega(func)}{block} = \frac{op_\omega}{block} \cdot \mu(func|\omega) \ .$$

This includes only the operation count for the function and not the operation count for functions which are called by this function (unlike programs such as *prof* or *gprof* where the processing time of a call tree is displayed). Counting only the operations in the function itself makes it easier to find the hot spots of the program in the source text. The total operation count is given by summing over all functions

$$\frac{op_\omega}{block} = \sum_{func \in F} \frac{op_\omega(func)}{block} \ .$$

The functions are sorted according to their operation count by introducing a rank function

$$rank : F \to \{0, \ldots, |F| - 1\}$$

such that

$$rank(func_1) < rank(func_2) \quad \Longleftrightarrow \quad \frac{op(func_1)}{block} \geq \frac{op(func_1)}{block} \ ,$$

where $func_1, func_2 \in F$. This means, the rank function numbers the functions in the program according to their operation count. The most important function gets label 0. In case of disambiguity, the order of the functions in the program is used in the numbering. A set of important function IF can be defined by selecting the first few functions $n_{important} \ll |F|$ from the ordered set of functions

$$IF = \{func \in F : rank(func) < n_{important}\} \ .$$

In case of Table 4.1, $n_{important} = 3$ would be sufficient. These functions might be manually optimized by changing the algorithms or by improving the software implementation. Alternatively they can be distributed among several processors. Furthermore, processors like the ARM7TDMI [174] offer the possibility to change the instruction set: a denser but slower instruction set can be used for all but the important functions. This achieves high performance while keeping the opcode sizes small. The effect of concentrating on the important functions can be illustrated by looking again at the small example. The improvement is quantified by the Performance-Cost Gain PCG for the most important functions

$$PCG(n_{important}) \quad = \quad \frac{\frac{Perf_{improve}}{ImprovementCosts + cost_{init}}}{\frac{Perf_{init}}{cost_{init}}} \ ,$$

$$where \quad ImprovementCosts \quad = \quad n_{important} \cdot cost_{improve}$$
$$Perf_{improve} \quad = \quad Perf_{init}(1 - \mu(IF) + \delta \cdot \mu(IF)) \ .$$

$cost_{improve}$ are the costs for improving a function. It is assumed that these costs are proportional to the number of functions which are improved, i.e., recoding parts of the functions will take equal amounts of time. In the example it is assumed that each improvement takes about 10% of the initial design cost ($cost_{init}$). δ denotes the improvement that is achieved for the function. In this example it is assumed that the performance of the function can be doubled when improving it. Using the data of Table 4.1 results in a factor of

$$PCG(n_{important} = 3) \quad = \quad \frac{1 - \mu(IF) + 2\mu(IF)}{n_{important} \cdot \frac{cost_{improve}}{cost_{init}} + 1}$$

$$= \frac{1 + 0.96}{3 \cdot 0.1 + 1}$$
$$= 1.51$$

performance-cost gain when improving the first three functions. The complete program consists of $|F| = 150$ functions. Improving all these functions gives a *PCG* of only 0.125. This means the overall performance-cost ratio decreases. There is only a marginal improvement of the performance. But this requires a significant increase of the design costs. The return for the design time is diminishing. This is one of the crucial aspects when designing complex systems. It is essential to make a careful analysis before trying to improve certain parts. In small systems there might be not such a big difference between improving all functions and improving just the important functions. But as the systems get more complex, the analysis gets more important. In the worst case, it is possible to reduce the overall performance / cost ratio. This occurs if the design time is wasted on useless optimizations. A second aspect of the *PCG* is the ratio of the initial costs versus the improvement costs. If the initial design costs are high, e.g., a microprocessor developed with high manual effort, the costs for the additional improvements will be justified. The overall design costs are not increased very much by adding the improvement costs. The other extreme is a specialized embedded system. It will be developed with low budget and a high degree of design automation. In this case the initial design costs are comparatively low. Hence, the designer must be careful when adding the improvement costs. Without a thorough analysis, the optimization can even reduce the competitiveness of the product.

Decreasing the granularity of the analysis from the function level down to the basic block level makes it very hardware dependent. The average basic block size for the video compression programs is in the order of 5 ... 10 operations. This is well in the size of a normal processor pipeline. This means, the optimizations on this level must take into account the hardware details of the particular processor which is not desirable in the early stages of the design. But the operation count per function is not detailed enough to permit the optimization described in the IDCT example at the beginning of this chapter. Therefore the operation count is refined along a second axis.

Instruction Mix

The second axis of refinement is the *instruction mix*. It addresses the occurrence probability of operations from a given instruction-set. The instruction set, i.e., the set of permissible operation types, is denoted by T. This is the pool of instruction which the compiler might use to translate the program from a high-level language, like C, into a processor dependent format. The processor can implement the operation types of the instruction in different ways. It can provide functional units which directly perform the operation (often used in RISC processors) or it can process a sequence of these basic operations which are stored in a micro-program (this is used for complex operations in CISC processors). In this thesis, the term *operation* is used to denote a simple three operand RISC operation. The term *instruction* is used for the complete instruction word of the VLIW processor. An instruction might consist of several operations. For a scalar or superscalar processor the terms operation and instruction are interchangeable.

The instruction mix is the probability of using a certain operation type. Again, this is added as a new dimension to the probability space $(\Omega \times F \times T, \mathbf{B}, \mu)$. For notational convenience, the instruction mix is denote as a diagonal matrix **InstMix**

$$\mathbf{InstMix}_\omega \quad = \quad \mathbf{diag}(InstMix_\omega(type_0), ..., InstMix_\omega(type_{|T|-1}))$$

$$= \begin{pmatrix} InstMix_\omega(type_0) & \cdots & 0 \\ 0 & \cdots & 0 \\ \vdots & \ddots & \vdots \\ 0 & \cdots & InstMix_\omega(type_{|T|-1}) \end{pmatrix}$$

where

$$InstMix_\omega(type_i) \quad = \quad \mu(type_i|\omega) \ .$$

As an example, we can consider the data from Table 4.1. In this case, the instruction-set consists of

$$T = \{ADD, SUB, MULT, ...\} \ .$$

$type_0$ is the ADD operation. ω is the *moglie.mpg* sequence. We might restrict the analysis to the *j_rev_dct* function, i.e., $F \equiv \{j\_rev\_DCT\}$. In this case we have

$$InstMix(type_0) = InstMix(ADD) = 33.88\% \ .$$

The instruction mix can depend on the input data sequence. A sufficient ensemble of input sequences has to be analyzed. As the results of section 7 will show, there is only little variation of the instruction mix when comparing different image sequences. So we can use, for example, the average instruction mix distribution as a representative data set for the system design.

The instruction mix allows to refine the design equation 4.2. It takes into account the different operation types

$$OPC(type) = \frac{InstMix(type)}{util(type)} \cdot \frac{op}{block} \cdot \frac{blockRate}{f_{Clk}} \ ; \qquad type \in T \ . \qquad (4.3)$$

This is important for the hardware design where different functional units will be used to implement the operation types. The instruction mix allows to calculate how many operations of a certain type are required to process a certain data block. Formally, the first two terms of equation 4.3, i.e., $InstMix(type)$ times $\frac{op}{block}$. This is the required number of operations. The $util(type)$ is necessary, because the processor must typically provide more than the required number of operations. There might be idle cycles due to data dependencies or resource conflicts. Thus only a fraction of the peak performance $OPC(type)$ is used for the processing. This fraction is defined by the utilization, i.e., $util(type)$. The last term $\frac{blockRate}{f_{Clk}}$ will be considered as a constant during the architectural development. The $blockRate$ is defined by the frame rate and the frame format that should be processed. This is the desired performance level of the embedded system. As mentioned above, it is typically defined at the beginning of the design. The hardware clock rate f_{Clk} describes the semiconductor and circuit technology. All processing times are normalized in terms of clock cycles. This allows to compare architectural improvements across different circuit technologies. The designer has to keep in mind that architectural decisions can influence the clock rate. For example, a complex operation might reduce the pathlength $\frac{op}{block}$, but it can reduce the clock rate as well. In this way the architectural improvements would be canceled by the slower clock speed of the final implementation.

For the design of the video processor it is assumed that the clock period is roughly equal to the delay of a 32-*bit* addition plus the delay of the latches for holding the results and the input operands. This allows to execute arithmetic RISC instructions in a single clock cycle. It can be regarded as an reasonable assumption about the final hardware design. The assumption is used to specify the second unknown of equation 4.3: f_{Clk}. All processing times are expressed in terms of clock cycles.

Equation 4.3 can be used to answer several design questions on a very abstract level. First of all, we can assume an ideal situation: an application is given with little data dependencies and an ideal scheduler which always finds a schedule without inserting *nops* due to data dependencies. In other words, the data dependencies are neglected for the time being. What should the data path of the processor look like in this ideal case? Idle cycles can only occur due to resource conflicts. Idle cycles decrease the overall performance-cost ratio because they do not contribute to the performance, but the system provides resources in the idle times. These idle resources increase the system costs. However, they do not contribute to the useful operations. Avoiding idle cycles means to avoid resource conflicts. Therefore the processor should provide the ideal data path for a given application, i.e., a processor should be derived that achieves 100% utilization for all operation types

$$\mathbf{util} = \mathbf{I} \; ,$$

where $\mathbf{I}$ denotes the identity matrix. Solving for the $\mathbf{OPC}$ that should be provided by the processor gives

$$\mathbf{OPC} = c \cdot \mathbf{InstMix} \; , \qquad (4.4)$$

$$where \quad c \equiv \frac{op}{block} \cdot \frac{blockRate}{f_{Clk}} \; .$$

In other words, the application and the available semiconductor technology demand a certain throughput requirement c in terms of operations per cycle. The distribution of these operations among the possible operation types is defined by the instruction mix. Thus multiplying the required throughput c with the probability of using a certain operation type $InstMix(type)$ gives the number of operations the processor should provide. In the ideal case the processor might even provide fractional number of operations, i.e., $OPC(type) \in \mathcal{R}$. In practice this can be achieved by using multi-cycle operations or different system clocks for the functional units. However, in this thesis the fractional operation numbers are mainly used as abstract quantities for calculating the processor structure. As an example of the $\mathbf{OPC}$ calculation, we assume an instruction set $T_{example} = \{add, mult\}$. An instruction mix

$$\mathbf{InstMix}_A = \mathbf{diag}(75\% \; 25\%)$$

and a throughput requirement $c_A = 4\frac{op}{cycle}$ are given. The processor should provide an

$$\mathbf{OPC} = 4 \cdot \mathbf{diag}(75\% \; 25\%) = \mathbf{diag}(3\frac{add}{cycle} \; 1\frac{mult}{cycle}) \; . \qquad (4.5)$$

This might be achieved by using 3 adder and 1 multiplier. The designer has several options to change the instruction mix distribution. Operations might be clustered into multi-functional units. For example, a second instruction mix might be considered where adds, subtracts, and logic operations are all used with a probability of 30%. A throughput requirement of $1\frac{op}{cycle}$ is given, the three operation types can be implemented by a single multi-functional ALU.

Implementing the same program on a processor with functional units that implement only one operation type is less efficient. In that case, the processor would need three functional units (adder, subtracter, and logic unit). Each of the units is used only during 30% of the time. Thus, most of the processor resources remain idle, making the design inefficient. However, this statement interferes with the throughput requirement. If the throughput is increased from $1\frac{op}{cycle}$ to $3\frac{op}{cycle}$, the situation changes completely. Now, the utilization of the three specialized units is as good as the utilization of the three ALUs. But the specialized units are smaller than the multi-functional units. Therefore, the specialized approach is better for the higher throughput system.

The discussion shows the complexity of embedded system design. Often design decisions depend on a multitude of different factors. In particular, the specific operating conditions have to be considered. The analysis data are indispensable because they allow to quantify the effect of certain design decisions. But without the designer's knowledge, they are next to useless. This becomes even more apparent when looking at the real design context. In that case, the designer has even more possibilities than just changing the hardware. He/she might chose to change the important functions. The instruction mix can be changed so that the resulting OPC contains integers. This can even improve the performance if the overall operation count is increased (see below).

Up to this point only a single application was considered. But as outlined by the multimedia examples of chapter 2, this application domain requires several different applications to run on the same hardware. What utilization can be achieved in this case? As an example, we can consider a second application B that is mapped onto a processor designed according to equation 4.5. The instruction mix of the second application is given by

$$\mathbf{InstMix}_B = \mathbf{diag}(25\% \; 75\%) \; .$$

In this case a throughput of $c_B = 4\frac{op}{cycle}$ cannot be achieved. This would require to execute 3 multiplications per cycle, but the hardware of equation 4.5 provides just a single multiplication per cycle. The multiplier is the bottleneck in this case. Application B would use the multiplier all the time and one of the three adders every third cycle, i.e., the utilization of application B is given by

$$\mathbf{util}_B = \mathbf{diag}\left(\frac{1}{9} \; 1\right) \; .$$

The overall throughput of application B is limited to just

$$c_B = \frac{1}{3} \cdot \frac{add}{cycle} + 1 \cdot \frac{mult}{cycle} = 1.33\frac{op}{cycle} \; ,$$

i.e., 33% percent the throughput of application A. The designer has two major possibilities to improve this situation, either the processor is designed in such a way that it gives moderate performance for both applications (this would result in a general-purpose processor) or the designer uses applications which fit to each other and the processor is adapted to this narrow application domain. The latter case is used for the embedded system design. In this case a condition is necessary that defines *similar* applications more precisely. It can be derived from equation 4.3. Matrix notation is used to simplify the notation

$$\mathbf{OPC} \;=\; c \cdot \mathbf{util}^{-1} \cdot \mathbf{InstMix} , \tag{4.6}$$

$$where \quad c \;\equiv\; \frac{op}{block} \cdot \frac{blockRate}{f_{Clk}} .$$

Adapting the processor to the first application A, as in equation 4.5, defines the processor structure by

$$\mathbf{OPC} = c_A \cdot \mathbf{InstMix}_A .$$

The units of the data path are chosen to match the instruction mix of application A. Mapping a second application B onto this processor means we get the following utilization for the second application

$$\begin{aligned}
\mathbf{util}_B \;&=\; c_B \cdot \mathbf{OPC}^{-1} \cdot \mathbf{InstMix}_B \\
&=\; \frac{1}{\alpha} \cdot \mathbf{InstMix}_A^{-1} \cdot \mathbf{InstMix}_B ,
\end{aligned} \tag{4.7}$$

$$where$$
$$\alpha \;\equiv\; \frac{c_A}{c_B} .$$

In other words, we take the data path structure that was optimized for the first application. This structures is used to process the second application and the resulting utilization is determined. α is the total degree of performance reduction that occurs for the second application. It can be determined by noting that the utilization for application B cannot become large than 100% for each operation type

$$\forall type \in T \;:\; 0 < util(type_i) \leq 1 .$$

Since data dependencies are neglected and fractional numbers of operations per cycle are permitted, at least one of the units will achieve 100% utilization (in the example above the multiplier). This operation type determines the throughput

$$\begin{aligned}
\exists\, type \in T \;&:\; util(type) = 1 \\
\Longrightarrow \;\; \alpha \;&=\; \frac{InstMix_B(type)}{InstMix_A(type)} .
\end{aligned}$$

This means, the bottleneck unit will slow down all other units until they fit to the maximum throughput of the bottleneck unit. More specifically, α is determined by the operation type where the largest difference between the two instruction set occurs

$$\alpha = \max_{type \in T} \left\{ \frac{InstMix_B(type)}{InstMix_A(type)} \right\} . \qquad (4.8)$$

In other words, equation 4.8 provides a procedure for quantifying the reduction of throughput that occurs if two applications are mapped onto the same hardware. In this case the instruction mix for both applications must be measured. Next, the instruction mix ratio is determined for all different operation types. The maximum ratio defines the reduction of the throughput.

In summary, both applications achieve only a high utilization if their instruction mixes are similar. In the ideal case they would have exactly the same instruction mix. This would allow to achieve 100% utilization for the second application as well. The designer can support this by clustering operation types so the resulting instruction mixes agree as much as possible. Alternatively, the most important functions might be recoded to obtain similar instructions mixes. For example, a fast IDCT can be replaced by a matrix formulation to make the instruction mix similar to the instruction mix of an FIR filter. In this case the filter and the IDCT will both achieve good utilization of the processor. But this requires again a careful trade-off because the matrix multiplication might increase the pathlength $\frac{op}{block}$. This means, the designer must balance the reduction of throughput, that occurs due to the mismatch of the instruction mixes, against the possible increase in the pathlength. The following problem must be solved. Given a processor that is adapted to an application A and a second application with two alternative implementations B and B', which of the two applications should be selected, i.e., which of the two implementations can achieve the highest $blockRate$? This can be answered by solving equation 4.7 for the block rate of the second application

$$blockRate_B = \frac{c_A \cdot f_{Clk}}{\frac{op_B}{block} \cdot \max_{type \in T} \left\{ \frac{InstMix_B(type)}{InstMix_A(type)} \right\}} . \qquad (4.9)$$

Comparing the block rates of the two alternative implementations gives the condition for selecting implementation B

$$blockRate_B \quad \geq \quad blockRate_{B'}$$

$$\frac{op_B}{block} \cdot \max_{type \in T} \left\{ \frac{InstMix_B(type)}{InstMix_A(type)} \right\} \leq \frac{op_{B'}}{block} \cdot \max_{type \in T} \left\{ \frac{InstMix_{B'}(type)}{InstMix_A(type)} \right\}$$

The implementation where the product of pathlength times throughput reductions is smaller, is the one which achieves the better block rate. The combined effect of possible pathlength increase and throughput reduction, caused by the instruction set mismatch, must be considered. Therefore the $\frac{op}{block}$ and the instruction mix for both candidate implementations must be measured. Next, the throughput reduction due to the mismatch with application A is calculated for both implementations and, finally, the implementation is selected which achieves the smallest product of pathlength increase times throughput reduction.

Cache Size and Organization

So far the requirement analysis has focused on the combined improvement of algorithms and software implementations for a spectrum of related tasks. Additional data like branch statistics and cache analysis data can be used in the same way. The measured data for the video applications are discussed in section 7. But the requirement analysis is not only important for improving the application programs. It is also used for restricting the design space. This is important since the next analysis step, the design exploration, is extremely demanding in terms of processing power. Therefore the results of the requirement analysis are used to focus the design exploration on the most important design domain. Again, knowledge about the application domain and many different aspects of the system design have to be considered. This will be exemplified by discussing results of a cache analysis. An example of a cache analysis is shown in Fig. 4.3. The *mpeg_play* program was analyzed and the *tennis.mpg* sequence was used as data input. The x-axis shows the line (or block) size of the cache. The y-axis shows the miss ratio that is achieved when changing the line sizes. Different cache sizes are used as parameters. A direct mapped cache is shown. In this case, the least significant bits of the address directly determine the cache line [162]. In this cache structure only one tag must be checked to determine if the selected data are cached or not. It makes the cache fast and avoids the associative memory for storing the tags. On the other hand it can increase the number of conflicts if data with the same lower address bits compete for a single cache line. The demanded cache line will be frequently replaced which leads to a higher cache miss ratio. A set associative cache or a fully-associative cache will outperform the direct mapped cache under these conditions.

Here, the direct mapped cache configuration is mainly used as a worst-case scenario to evaluate the maximum memory traffic that can occur in the embedded system. Typically, the design of the memory system is one of the major cost factors for a modern computer system [76]. Hence, it is important to restrict the design exploration in such a way that the number of load / store units is in a useful range. An estimate on the memory traffic is obtained by combining the data from the instruction mix analysis and the cache analysis.

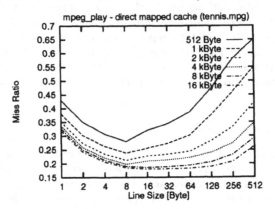

Figure 4.3. Cache analysis data for the *mpeg_play* program and the *tennis.mpg* data sequence.

We assume an $\frac{op}{block}$ of $10k$ operations for the *mpeg_play* program. Furthermore, we assume that approximately 30% of these operations are load / store operations. These settings reflect closely the results obtained for the compression system. The details are presented in chapter 7. The ideal number of load / store units can be calculated by equation 4.4. A clock rate of $f_{Clk} = 50MHz$ and a real-time processing of the CCIR601 format (see Table 2.1) is assumed. This means, a data block must be processed in

$$\frac{50MHz}{40.5kBlocks/s} = 1234.56 \cdot \frac{cycles}{block} .$$

The OPC for the load/store units is given by

$$OPC(load/store) = InstMix(load/store) \cdot \frac{op}{block} \cdot \frac{blockRate}{f_{Clk}}$$

$$= 0.3 \cdot 10k \cdot \frac{40.5k}{50M} \frac{op}{cycle}$$

$$= 2.43 \frac{op}{cycle} .$$

As a result, approximately 3 load / store operations will occur in each cycle. Such a high memory traffic is typical for video compression applications [149]. The cache acts as a kind of filter which answers directly all the memory requests concerning cached data. If the data are not stored in the cache, a miss occurs. A line of the cache will be replaced by a line from the memory that contains the requested data. The cache configuration must be selected before the miss ratio can be determined. Many different design aspects have to be taken into account [162]. Increasing the line size of the cache will reduce the access time of the cache because fewer tags have to be stored. But a large line size will also require more memory traffic in case of a replacement. The data in Fig. 4.3 show that a line size of 8$Byte$ is the most appropriate one. This agrees with the 8 pixel vector that is frequently used in the compression algorithms. A smaller line sizes increase the miss ratio because the compression algorithms often access a row of 8 data bytes. A smaller line size produces a miss in the middle of such a row. With a line size of 8$Byte$, the complete vector is loaded when the first miss occurs. Using just a line size of 1$Byte$ will produce a cache miss for every element of the vector, i.e., it will increase the overall miss ratio. Such a small line size would only be useful if the algorithms operate mainly on data which are scattered in memory. Increasing the size of the cache tends to lower the miss ratio. However, it will also increase the access time of the cache. In this case, increasing the size from 8$kByte$ to 16$kByte$ will not improve the performance of the cache anymore. This can be explained by the processing of data streams in the video compression algorithms. There is no stable working set which can be kept in the cache. Instead, each processing of a new macroblock will require new data in the cache. Only the temporary data that are used during the processing of a data block profit from the caching. There is no benefit in providing more cache space than required for the temporary data.

To estimate the memory traffic, a line size of 8$Byte$ is assumed and a worst-case miss ratio of 0.3. This means approximately one replacement will occur in every cycle. In the system of Fig. 2.4, this replacement will be done by the static memory. In most cases the memory is slower than the processor. An access might cost 2 cycles. The memory bandwidth would not be sufficient. This can be remedied by performing several memory operations in parallel. A wider data bus will be used in this case. The bus increases the pin-count of the compression processor which is another major cost factor. An estimate of the data bus width is given by

$$busWidth = (OPC(load) + OPC(store)) \cdot missRatio \cdot lineSize \cdot \frac{cycles}{access}$$

$$= 3\frac{op}{cycle} \cdot 0.3 \cdot 8Byte \cdot 2\frac{cycle}{op}$$

$$= \ 115.5bit \ .$$

A data bus of 128-*bit* must be used. Using more than 3 load / store units would require an even larger bus. A more expensive package has to be used. This would increase the system costs significantly. As a consequence, a higher number of load / store units in the processor's data path would only be justified if a significant performance gain can be achieved. The design space of processors with more than 3 load /store units needs not so much attention in the exploration. Unless, it is shown that any dramatic performance improvements can be made.

This discussion shows how many different trade-offs have to be considered when making design decisions on the system-level. A large number of different design constraints and possible alternatives has to be taken into account. It is not enough, to use worst-case assumptions in all cases. This would lead to completely unrealistic design which is oversized for most cases and thus not competitive. The main challenge is an estimation of a realistic system structure without implementing it in detail. The basis for this are the design data which are measured on an abstract level. The most important data of the requirement analysis are the $\frac{op}{block}$ and the **InstMix**. In addition, the cache analysis and branch statistics are necessary. The next section refines the analysis so that it takes into account the data dependencies among the operations.

DESIGN EXPLORATION

So far the analysis has determined a lower bound on the peak performance that is required to process the video compression applications in real-time. It was investigated how the programs can be improved by recoding the most important functions. Conditions were derived for mapping several applications onto the same processor structure. Throughout this analysis the data dependencies were neglected. An ideal scheduler was assumed which achieves a utilization of 100%. Only idle times due to instruction mix mismatch were considered. This section refines the data by taking into account the scheduling inefficiencies and the resource conflicts. The VLIW processor framework of section 2.4 is used as the basis of the exploration. Data dependencies hinder parallelization because one operation uses the result produced by another operation. The consuming operation has to wait until the other operation has finished the calculation of the result. Both operations cannot be executed in parallel, even if a sufficient number of functional units is available. Resource conflicts are different. They occur because the processor has not enough functional units. The operations have to compete for the functional units. In principal, they can execute in parallel, but the processor has not enough hardware resources to exploit the parallelism.

The effect of data dependencies and resource conflicts can be seen by looking at two extreme processor structures. First, we can consider a scalar processor with a single, multi-functional unit and a small register file. In this case the processor cannot execute operations in parallel. Therefore resource conflicts appear whenever operations permit parallelism. The other extreme is a processor with an infinite (or very large) number of functional units. In this case, there will be no resource conflicts. So only data dependencies will hinder parallelization. Both processor structures are normally not optimal, because they provide a low performance / cost ratio. In the first case, the performance is low. A small increase in the number of functional units will typically increase the performance significantly (without increasing the costs very much). The second structure has extremely high system costs. Reducing the the number of functional units will give approximately the same performance, but at significantly lower costs.

If we increase the processor structure gradually, from a scalar processor to an infinitely large processor, we will see the following behavior of the performance / cost curve. It will start with a low value due to the low performance of the scalar processor. Next, the ratio will increase because the performance increases when adding more units. At the same time, the system costs increase only very slowly. A maximum value will be reached when almost no more resource conflicts occur *and* there are only few idle times due to data dependencies. Finally, the performance / cost ratio will drop again, because adding more units increases the costs, but there are almost no more performance gains as a consequence of the data dependencies.

The system designer will, of course, look for the maximum of the performance / cost ratio. Unfortunately, the scenario above is very much idealized. In particular, it is difficult to assess the costs. This was already discussed in section 4.2 when designing the cache for the system. In that case, the packaging costs had a very non-linear effect.

Additionally, there are mandatory performance levels, that have to be fulfilled to make the product competitive (decoding a certain compression standard for a specific frame format). These constraints must be met due to market conditions (they are almost psychological). This even applies if there are points which offer a better design trade-off. Moreover, the performance / cost ratio is not a curve which depends only on a single parameter, because the VLIW processor structure allows to change several parameters. The resulting performance will depend on all these parameters. There is no systematic way to "increase" the processor structure. Which unit type should be increased first? How much are the overall system costs effected by a larger number of units? Is it better to use fewer, more expensive units? Often the reuse of components changes the costs completely. For example, a full-custom design of a register file is very useful (see section 7.9). Reusing this design will often lead to a smaller circuit, in comparison to synthesizing a register file. In this case the existing file will be used, even if the size is to large for the current design. In summary, assessing all the influences on the system costs requires very much knowledge. Therefore it is not useful to generate the performance / cost ratio directly. The results would interfere too much with the assessment of the costs. The tools should provide only the basic data. The design exploration will determine only the performance for the different processor structures. The designer can use his/her knowledge about the complete context to assess the costs. At this point, the design exploration on the system-level differs from a design exploration on lower levels of abstraction [100] [40]. In the latter case, the exploration typically optimizes gate-count costs versus performance. But this is no more adequate on the system-level where a multitude of different influences have an effect on the costs.

The design exploration on the system-level determines the processing times (reciprocal of the performance) for processor variants in the current design context. The exploration approach is not limited to the VLIW architecture (although it is advantageous since a set of parameters can be used to alter the processor structure). Other processor architectures can be investigated in the same way. For example, Johnson's general model of a superscalar processor [103] can be used in the same way. In fact any architecture which provides a systematic way for producing processor variants can be explored.

The exploration poses two main problems. First of all, the number of possible processor structures is very large. It leads to extremely large processing times for the exploration. A reduction of the exploration complexity is necessary. The second problem is the access of the data. The designer must get a selective access to the important results. He/she cannot look at the results of all explored processors, rather, he/she will be interested only in the most suitable processor structures. A mechanism must be provided to focus on these parts of the exploration.

Definitions: Processor, Design Space, and Exploration

In the following, the problems will be addressed in detail. First, the reduction of the complexity will be discussed. Therefore it is necessary to define the exploration process more precisely. Basically, it determines the processing times for a set of processors from a certain design space. The processing time of a processor is determined in the following way. First, the programs are scheduled for this processor. Next, the cycle time for a basic block, in the scheduled code, are multiplied by the execution frequencies of the blocks. This gives the total processing time for that single basic block. Taking the sum over all basic blocks results in the processing time for a single application program on a single processor structure (the process is defined in detail below). Repeating this process for a number of different processor structures (the design space) gives the desired exploration data. But scheduling of all application programs for all possible processor structures would be prohibitive due to the required processing time. Therefore a stepwise approach is used. First of all, the resource constraints due to the interconnection network are excluded. The design of the interconnection network will be addressed during the data transfer analysis that follows the design exploration. Throughout the design exploration we assume relaxed processors where all functional units are completely connected, i.e., a bus is available for all possible transfers. This means, the scheduler will not postpone any operation due to a missing bus access. It is not useful to implement the relaxed processor because the interconnection network would be too large and slow, i.e., the clock rate of the processor will be very low and the real-estate required for the interconnection network would be extremely large. But scheduling for the relaxed processors is useful to derive bounds on the cycle time. The relaxed processors are vehicles for the exploration procedure. They are not meant to be realized. In the real processor there will be additional idle cycles due to resource conflicts for the busses. The cycle time for the real processor will at best be equal to the cycle time achieved by the relaxed processor. In typical cases it will be larger than the cycle time of the relaxed processor. In this way the cycle time of the relaxed processor is a lower bound on the cycle time for a real processor.

Excluding the interconnection network from the exploration is one possibility to reduce the complexity. The second possibility modifies the exploration strategy. For this purpose a more formal definition of the exploration is required. We start the derivation by defining a hardware library **UL**

$$\mathbf{UL} = (unit_0, unit_1, \ldots, unit_{|L|-1}) \,, unit \in L \,.$$

L denotes the set of available unit types, e.g., $L = \{adder, multiplier, \ldots\}$. These are the units that can be used to configure the data path of the VLIW processor from Fig. 2.6. In principal, it is possible that a unit can implement several operation types. For example, an ALU might implement addition, subtraction, and logic operations. As a consequence, it would be difficult to compare the results of the requirement analysis with the results obtained during the design exploration. Therefore we assume that each $unit_i$ implements exactly one operation $type_i$. **UL** arranges these unit types as a vector. Each dimension of the vector **UL** corresponds to one unit type. A relaxed processor is defined by selecting the number of functional units, i.e., a function

$$P \; : L \to \mathcal{Z}_+ \cup \{\infty\}$$

is defined where $\mathcal{Z}_+ \equiv \{0, 1, 2, \ldots\}$. P is a function that returns the number of units for each unit type. There are $P(unit_i)$ elements of $unit_i$ in the processor. The complete selection of unit types can be considered as a point $\mathbf{P}$ in a design space $\mathbf{DS}$. The elements of the vector $\mathbf{P}$ are arranged as defined by $\mathbf{UL}$, i.e.,

$$\mathbf{P} = (P(unit_0), P(unit_1), \ldots, P(unit_{|L|-1})) .$$

The design space $\mathbf{DS} \subset (\mathcal{Z}_+ \cup \{\infty\})^{|L|}$ is defined by

$$\mathbf{DS} = DS_0 \times DS_1 \times \ldots \times DS_{|L|-1} . \tag{4.10}$$

Each of the subspaces DS_i denotes a set of possible unit numbers for a unit type $(unit_i)$, e.g., $DS_i = \{1, 2, \ldots, 6\}$. A processor belongs to this design space if all selected unit numbers are in the permitted ranges

$$\mathbf{P} \quad \in \quad \mathbf{DS}$$

$$\Longleftrightarrow \quad \forall i \in \{0, \ldots, |L| - 1\} \quad : \quad P(unit_i) \in DS_i .$$

As an example, we can state the definitions for the processor that was used in the instruction mix example of section 4.2. In this case the set of unit types is $L_{example} = \{adder, multiplier\}$. The unit library is defined as

$$\mathbf{UL}_{example} = (adder \; multiplier) .$$

The example processor is given by

$$\mathbf{P}_{example} = (3 \quad 1) ,$$

It has 3 adders and 1 multiplier. It belongs, for example, to the following design space

$$\mathbf{P}_{example} \in \mathbf{DS}_{example} = \{1, \ldots, 4\} \times \{1, 2\} .$$

This design space consists of all processors that have 1 to 4 adders and 1 or 2 multipliers. There is a total of 8 possible processor structures in this design space.

Now it is possible, to define the design exploration itself. The exploration starts by selecting a design space $\mathbf{DS}$. For each processor in the design space, $\mathbf{P} \in \mathbf{DS}$, the application program is scheduled. Next, the cycle time

$$\frac{cycle_\omega(\mathbf{P})}{block} \quad : \quad \Omega \times \mathbf{DS} \to \mathcal{R}$$

is determined. It denotes the processing time for the data sequence ω on the processor $\mathbf{P}$. It needs not to be an integer number due to the normalization on data blocks. The cycle time takes the following effects into account: (1) the data dependencies among the operations, (2) the resource conflicts that occur due to restricted numbers of functional units, and (3) the schedulers ability to arrange operations in a sophisticated way to avoid idle cycles [80]. For example, delay slots of jumps might be filled with operations preceding a jump. In this way waiting time for the outcome of a branch is used to perform other operations. In most cases the dependencies will reduce the utilization of the processor. If, for example, the three additions of the instruction mix example from section 4.2 depend on the outcome of the multiplication, the processing of the operations would take at least 2 cycles which reduces the average utilization to 50% for both unit types.

Peak Performance

A first step before starting the exploration is to look at the theoretical peak performance for a certain application. What performance can be achieved if a particular processor structure is used to process an application with a given instruction mix? In a more formal way this can be derived by combining the processor definition of this section with the design equation 4.6 of section 4.2. We start by noting that the processor structure defines integer numbers for each operation type, i.e.,

$$\mathbf{OPC} = \mathbf{diag}(\mathbf{P}) \; .$$

The fractional OPC numbers, that where permitted in the requirement analysis, are mainly an aid for the calculation. For real processor structures it is necessary to use integer numbers.

A one-to-one mapping between the operation types and the units is assumed, as discussed above. The peak throughput c_{peak} of this processor is given by

$$c_{peak} \;=\; \mathbf{P} \cdot \mathbf{1}^T = \|\mathbf{P}\|$$

$$where$$

$$\|\mathbf{P}\| \;=\; \sum_{unit \in L} |P(unit)| \; .$$

This means that the maximum number of operations per cycle which a processor can provide, is given by c_{peak}. It is equal to the number of functional units since each unit can execute one operation per cycle (multi-cycle operations are not permitted in the template of section 2.4). The instruction mix **InstMix**$_{peak}$ of an imaginary application, which would achieve the peak performance, can be determined. This is given by solving equation 4.4 for the peak instruction mix

$$\mathbf{InstMix}_{peak} = \frac{1}{\|\mathbf{P}\|} \cdot \mathbf{diag}(\mathbf{P}) \; . \tag{4.11}$$

The $\mathbf{InstMix}_{peak}$ can be considered as a direction in the design space. For example, if we double the number of functional units for each unit type, we obtain a processor with exactly the same peak instruction mix but with doubled peak throughput. The next step is to determine the throughput reduction that occurs when the real application is mapped onto the processor. First of all, the data dependencies can be neglected to derive the ideal throughput reduction α_{ideal}

$$\alpha_{ideal}(\mathbf{P}, \mathbf{InstMix}) = \max_{type \in T} \left\{ \frac{\|\mathbf{P}\| \cdot InstMix(type)}{P(type)} \right\} . \qquad (4.12)$$

This is the minimal reduction of throughput. It even occurs, if we imagine an ideal scheduler that can find a sophisticated arrangement of operations such that no data dependencies will ever stall the processor. The set of ideal processors in the design space $\mathbf{DS}$ is given by

$$\alpha_{ideal}(\mathbf{DS}, \mathbf{InstMix}) = \min_{\mathbf{P} \in \mathbf{DS}} \left\{ \alpha_{ideal}(\mathbf{P}, \mathbf{InstMix}) \right\} .$$

This means, there might be more than one processor with the same ideal throughput reduction. These processors can use all their resources with a minimum number of idle cycles, provided that the scheduler finds a schedule which avoids stalls due to data dependencies. The utilization under these ideal circumstances is given by

$$\mathbf{util}_{ideal} = \frac{\|\mathbf{P}\|}{\alpha_{ideal}(\mathbf{P}, \mathbf{InstMix})} \cdot \mathbf{diag}(\mathbf{P})^{-1} \cdot \mathbf{InstMix} .$$

Table 4.2 shows the throughput reduction of the example design space, $\mathbf{DS}_{example}$, if an application with an instruction mix $\mathbf{InstMix} = (0.75 \quad 0.25)$ is mapped onto the processors. Processor number 3 is the best suited processor in this case.

In the real world, a larger throughput reduction will occur due to the data dependencies among the operations. In this case the scheduler adds idle cycles to wait for the results of operations, i.e., the number of operations that the hardware provides is given by

$$\|\mathbf{P}\| \cdot \frac{cycle(\mathbf{P})}{block} = \frac{op}{block} + \frac{idle}{block} .$$

Another source of idle cycles, besides the data dependencies, is the synchronization with the frames. If a processor can process an image frame faster than required for the current frame rate, it has to wait until the next frame arrives (it is synchronized after processing each frame). In the following the idle times

Table 4.2.   Throughput reductions for the example processor design space.

Throughput Reduction in the Ideal Case

Number	P			Peak Inst Mix			
	adder	multiplier	‖P‖	add	mult	Reduction	Ideal Perf.
#1	1.00	1.00	2.00	0.50	0.50	1.50	1.33
#2	2.00	1.00	3.00	0.67	0.33	1.13	2.67
#3	3.00	1.00	4.00	0.75	0.25	1.00	4.00
#4	4.00	1.00	5.00	0.80	0.20	1.25	4.00
#5	1.00	2.00	3.00	0.33	0.67	2.25	1.33
#6	2.00	2.00	4.00	0.50	0.50	1.50	2.67
#7	3.00	2.00	5.00	0.60	0.40	1.25	4.00
#8	4.00	2.00	6.00	0.67	0.33	1.13	5.33

due to frame synchronization will be neglected. It is assumed the processor operates as fast as it can on the given image stream. In this case the block rate is equal to

$$blockRate = \frac{f_{Clk}}{\frac{cycle(\mathbf{P})}{block}} \ .$$

The utilization can be determined by the design equation

$$\mathbf{util} \ = \ s \cdot \mathbf{InstMix} \cdot \mathbf{diag(P)}^{-1} \qquad (4.13)$$

$$where \quad s \ \equiv \ \frac{\frac{op}{block}}{\|\mathbf{P}\| \cdot \frac{cycle(\mathbf{P})}{block}} \ .$$

s is the ratio between the number of operations that the application demands, after compilation, and the number of cycles the processor needs to execute for the application. For a scalar processor that can execute every operation in a single cycle, s is exactly 1. For a processor which executes several operations in parallel, this is the effective degree of instruction level parallelism that is achieved in each cycle.

Structured Exploration

The analysis performed so far allows to devise a first approach for a design exploration strategy. The strategy reduces the number of exploration points without neglecting parts of the design space. We start by selecting a design space. For each unit we select the upper bound on the unit range by using the maximum system costs of Fig. 4.1. Afterwards, the requirement analysis is performed. The throughput c is determined which is necessary to fulfill the real-time condition. Next, the α_{ideal} is calculated for every processor in the design space. The ideal throughput of each processor is determined by multiplying the peak throughput with $\frac{1}{\alpha_{ideal}}$. A subset of the original design space is formed consisting of all processors where the ideal throughput is above the throughput demand of the real-time condition. If a throughput requirement of $c = 4\frac{op}{cycle}$ is given for the example, only the processors $\mathbf{DS'} = \{\#3, \#4, \#7, \#8\}$ need to be considered for further exploration. In principal, these processors can achieve the real-time processing of the video applications. However, idle times, due to data dependencies, might reduce the performance below the performance level required for the real-time processing. Therefore, the cycle times must be determined for all the remaining candidate processors. The time required for this step is proportional to the number of processors that remain in the design space

$$\mathbf{|DS|} = \prod_{0 \leq i < |L|} |DS_i| . \qquad (4.14)$$

In other words, the complexity of the design space is equal to the product of all unit exploration ranges.

A simple estimate shows that such a brute force strategy will not work in practice. For example, the library might contain just $|L| = 10$ unit types. Each of these unit types is explored in a range consisting of $|DS_i| = 10$ possible settings. This would lead to a complete design space of 10^{10} processors. Scheduling one of the video processing programs on one of these processors takes about 10 min on a SPARCstation 10. This means the calculation of the cycle times would require $1.9 \cdot 10^5$ years. Clearly, a more sophisticated exploration strategy is necessary. A stepwise approach is used which consists of the following three main parts:

1. The *disjoint* exploration. The design space is selected in a way that only one unit type is restricted while all other unit types are left unbounded. In this way the maximum performance is determined that a processor can achieve with the certain number of units. However, the reduction of performance due to the combined restriction of several different unit types is not considered.

2. The *combined* exploration. A small subset of the original design space is selected. The selection of the subset is based on the results obtained during the disjoint exploration. For this restricted subset of the design space, all possible processor constellations are investigated. This takes into account the performance reduction that results from a combined restriction of several unit types. For example, a large number of adders can be useless if only a small number of load / store units is provided. In this case the adders might achieve high performance but the small number of memory ports would not supply an adequate number of input operands. The adders would idle due

to missing input data. A large register file is assumed during this step. This means that all intermediate results are stored in the register file. Only the new input data are fetched from the memory.

3. The *register file* exploration. This step addresses the selection of an appropriate register file size. A large register file allows to reduce the traffic to the main memory. This means a large register file is desirable from the processor architectural point of view. However, a large register file would be slow, i.e., it would reduce the clock rate of the processor [76] (see section 4.9). This means the register file must be kept as small as possible to make it fast. The designer must balance the performance improvements due to the larger size against the performance losses due to the lower clock rate. The suitable register size depends very much on the constellation of the functional units in the data path of the processor. If a large number of functional units is used, it will increase the number of intermediate results that have to be stored and the number of ports that the register file must provide.

Since the last two steps depend very much on each other, it is possible to interchange them. This would be required if there are severe restrictions on the size of the register file. In this case it is unnecessary to explore data paths with many functional units since the small register file hinders a high degree of instruction-level parallelism. In the following, each exploration steps will be explained in more detail.

Disjoint Exploration

The starting point of the design exploration is the disjoint exploration. In this case the resource conflicts are evaluated for each operation type separately. The resource conflicts for unit type $unit_i \in L$ can be explored by selecting the following design space $DS_{disjoint,i}$

$$\mathbf{DS_{disjoint,i}} \equiv \{\infty\} \times \cdots \times DS_i \times \cdots \times \{\infty\} , \quad DS_i \subset Z_+ .$$

The number of units for type $unit_i$ is selected from the specified exploration range, i.e., $unit_i \in DS_i$. But it is assumed that for all other unit types there is an infinite number of resources. Providing an infinite number of units means that no resource conflict can occur since there is always a free resource available. Infinity means in this case, there must be more functional units than the maximum degree of instruction-level parallelism in the program. Selecting an infinite number for all resources would result in an ASAP (As Soon As Possible) schedule. In this case the cycle time $\frac{cycle_w(P)}{block}$ is equal to the average length of the critical path, i.e., the paths of the program are weighted with there execution probability. The resulting cycle time is the expectation value of the critical path for processing blocks in the image sequence ω. Selecting an infinite number of resources for all units except $unit_i$, results in a modified ASAP schedule where all operations execute as soon as possible except those of $unit_i$. The situation is exemplified by the example from the beginning of this section where three additions depend on the result of the multiplication. The results of the disjoint exploration for this simple example are shown in Fig. 4.4.

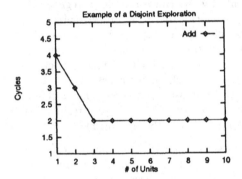

Figure 4.4.   Results of a disjoint exploration for a simple application. The application contains 3 additions which depend on a multiplication. Each operation takes one cycle to execute.

In this case the number of cycles is reduced from 4 cycles, if the application is processed with a single adder, to 2 cycles if three adders are provided. Using more than 3 adders does not improve the performance due to the data dependency of the additions on the result of the multiplication. The application reaches the bound defined by the ASAP schedule. This means in praxis, we do not need an infinite number of units to reach the ASAP schedule, three or more adders are sufficient. Resource conflicts are avoided, if the number of units is larger than the available parallelism for that unit type.

In the same way the disjoint exploration is carried out for the other units, i.e., a design space $\mathbf{DS_{disjoint}}$ is formed that consists of all these unit design spaces

$$\mathbf{DS_{disjoint}} = \bigcup_{0 \leq i < |L|} \mathbf{DS_{disjoint,i}} \ .$$

The complexity of the design space is given by

$$|\mathbf{DS_{disjoint}}| = \sum_{0 \leq i < |L|} |\mathbf{DS_{disjoint,i}}| = \sum_{0 \leq i < |L|} |DS_i| \ .$$

In the example from above this will reduce the processing time for the exploration from $1.9 \cdot 10^5$ years to

$$explTime = \frac{time}{processor} \cdot \sum_{0 \leq i < |L|} |DS_i| = 10min \cdot \sum_{0 \leq i < 10} 10 = 16.7 \ hours \ .$$

This means, the exploration complexity is reduced by several orders of magnitude. It is achieved without reducing the exploration ranges. The disjoint
exploration provides a complete overview of a large design space. The overview
is used to select the most important part of the design space. As an example,
we can take the results of Fig. 4.4. The performance changes significantly when
increasing the number of adders from 1 to 3. More than 3 adders gives no more
performance improvements. A processor with just 3 adders will achieve the
same performance as a processor with 5 or 6 adders. But the smaller processor
provides the performance at significantly lower costs. It is a typical situation
for most applications.

The instruction-level parallelism is often limited [104][116] due to data
dependencies. More detailed explorations, like the combined exploration, can
skip those parts where the performance is already saturated. It is very unlikely
that they will be used for the final design.

Combined Exploration

The design space of the combined exploration is therefore only a small subset
of the disjoint design space. The exploration range for each unit is the most
interesting part of the disjoint range

$$\mathbf{DS_{combined}} = DS'_0 \times DS'_1 \times \cdots \times DS'_{|L|-1} \ , \quad where \ DS'_i \subset DS_i \ .$$

Typically, $|DS'_i| \ll |DS_i|$ so that the overall exploration complexity is signifi-
cantly reduced.

A new problem is encountered for the combined exploration: the visualization of the exploration data. The exploration results form a multi-dimensional
data space which must be presented to the designer. This was not a problem
for the disjoint exploration. Each dimension was explored independently from
the other dimensions. Therefore, the data can be depicted as curves where
the range indices are used as x-values, the cycle count as y-value and the significant dimension as parameter (see Fig. 4.4). In the combined exploration,
the dimensions can no longer be treated independently from each other. For
example, the performance results of the adder depend on the number of load
units, etc. In other words, more than one dimension in the processor vector $\mathbf{P}$
is significant.

The situation will be exemplified by looking again at a small example. A unit library $\mathbf{UL}_{combined} = \{adder, multiplier\}$ is used. The following design space was selected

$$\mathbf{DS}_{combined} = \{1, \ldots, 4\} \times \{1, 2\} .$$

The design space consists of

$$|\mathbf{DS}_{combined}| = 4 \times 3 = 12$$

possible processors. An exploration was performed for the *PVRG-jpeg* [85] program. The *lena.jpg* sequence was used as data input. The results of this example exploration are shown in Fig. 4.5. The cycle times are normalized in terms of macroblocks. The effect of the combined restriction can be seen

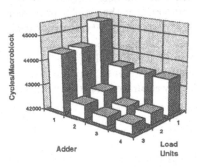

jpeg Example: Design Space Data			
Add	*Load Units*		
	1	*2*	*3*
1	45312.6	44397.1	44363.3
2	43570.6	42819.4	42573.4
3	43474.5	42458.5	42381.2
4	43472.5	42456.4	42379.2

(1) Graphic of the design space   **(2) Cycle times**

Figure 4.5.   The exploration example of the PVRG-jpeg with just 12 different processors.

by looking at all the processors with 4 adders. These are the three rightmost columns in the graph of Fig. 4.5. The cycle times with just one load unit are significantly lower than the cycle times with two or three load units. This means selecting a suitable number of adders cannot be done without considering the load units. For this simple example it can be determined by looking at the complete design space. In general, this means to look at an $|L|$-dimensional space which is not possible for $|L| > 2$. Design automation on lower levels of abstraction (logic synthesis and data path synthesis) often handles this situation by using a cost function and mathematical optimization techniques [100] [40]. This is more intricate on the system-level due to the variety of different influences that affect design decisions (as discussed above). Typically, these decisions are very difficult to quantify in a mathematical cost model. Knowledge of possible alternative implementations or market considerations is often involved. At certain points the selection of critical design parameters can change the complete design. As an example, we can recall the estimation of the memory bandwidth from section 4.2. If the design exploration shows that more memory ports are necessary than estimated, a more expensive packaging must be selected. Using this expensive package might in turn favor a system design for a high-end market. In this market, the higher system costs might be competitive. In other words, the results of the design exploration force the designer to reconsider a number of major design decisions which were made at earlier design steps. Such a global view would be completely out of scope for a mathematical optimization algorithm. The algorithm would just optimize the cost function, but cannot restructure the design. Therefore it is important in system-level design to provide the designer with a picture of the design space. The picture gives the designer a possibility to control the selection of the processor structure. Even complicated influences can be considered in this way.

In case of the multi-dimensional design space the picture is obtained by providing many different views. These views make it possible to look at the design space from many different perspectives (see Fig. 4.6). We start by defining the *unit view*. In this case the observer takes a view point on each of the axis in the design space. He/she looks at the design space from this perspective. More formally, a unit view for $unit_i \in L$ is a function

$$\mathbf{View_{unit_i, DS}} \ : \ DS_i \rightarrow \mathcal{R}^3 \ .$$

It generates a 3-tuple for every permitted unit number $P(unit_i) \in DS_i$, i.e.,

$$\mathbf{View_{unit_i, DS}}(P(unit_i)) = \begin{pmatrix} Min(unit_i, \mathbf{DS}, P(unit_i)) \\ Avg(unit_i, \mathbf{DS}, P(unit_i)) \\ Max(unit_i, \mathbf{DS}, P(unit_i)) \end{pmatrix} \ .$$

The values denote the minimum (Min), the average (Avg), and the maximum (Max) cycle time for all processors with $P(unit_i)$ functional units of type $unit_i$. The minimum cycle count is for example given by

$$Min(unit_i, \mathbf{DS}, P(unit_i)) \quad = \quad \min_{\tilde{\mathbf{P}} \in \widetilde{\mathbf{DS}}(P(unit_i))} \left\{ \frac{cycle(\tilde{\mathbf{P}})}{block} \right\} \ ,$$

$$where \quad \widetilde{\mathbf{DS}}(P(unit_i)) \quad \equiv \quad \{\tilde{\mathbf{P}} \in \mathbf{DS} : \tilde{P}(unit_i) = P(unit_i)\} \ .$$

It is the minimum cycle time for the processors in the restricted design space $\widetilde{\mathbf{DS}}(P(unit_i))$. This restricted design space is a subset of the original design space $\mathbf{DS}$. It is formed by selecting all processors with exactly $P(unit_i)$ units

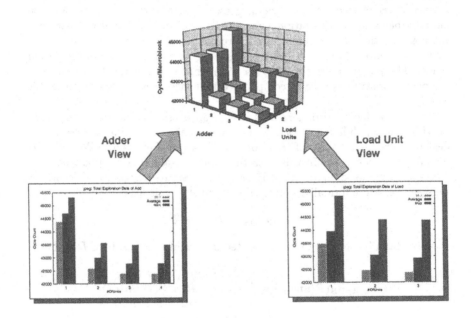

Figure 4.6.   Unit views on the design space of the *jpeg* example.

of type $unit_i$. Then the processing times on this restricted design space are searched to find the minimum cycle time $Min(unit_i, \mathbf{DS}, P(unit_i))$. The other elements are calculated in an analogous way.

In the *jpeg* example above, the 3-tuple for the adder view with $P(adder) = 1$ is found by looking at the first row of the table in Fig. 4.5. These are the processors

$$\widetilde{\mathbf{DS}}(P(adder) = 1) = \{(1\ 1), (1\ 2), (1\ 3)\} \ .$$

The corresponding minimum cycle time is given by

$$Min(adder, \mathbf{DS}, 1) = \{45312.6, 44397.1, 44363.3\} = 44363.3 \ ,$$

the minimum of the first row in the table of Fig. 4.5. The processor $\mathbf{P} = (1\ 3)$ achieves the minimum in this case.

The unit view is an aid to depict the consequences of what happens, when the number of units for a specific unit type is changed. Two questions need to be answered. First, what performance gains are achieved when increasing the number of units? This is answered by the average value. It shows the processing time that can be reached with a certain number of units. In the example of the adder view in Fig. 4.5, the performance that is achieved with 1 to 4 adders. But, as explained above, this performance depends on the constellation of the remaining unit types. The average value shows only, what performance is achieved for the "average" constellation. It means that the constellation is neither very well chosen nor extremely bad. This motivates the second question. How sensitive is the performance to the constellation of the other units? How independent is the selection of the current unit type from the selection of the other unit types? This is answered by the minimum and maximum value. If the maximum and minimum cycle time differ very much, this indicates that the performance is significantly influenced by the other units. For example, in the *jpeg* example, a single load unit is a bottleneck. This manifests itself in the variation of the minimum and the maximum. As an example, we can look at the performance of processors with 2 adders. The best cycle time is 42.5 k cycles. It is achieved with 3 load units. 2 load units achieve almost the same performance level. Whereas the worst-case performance is achieved when just a single load unit is used. It gives the maximum cycle time of 43.5 k cycles. The variation of the minimum and maximum cycle count indicates this interdependency. A very small variation means, the unit type is almost independent from the other unit types. This is shown by a subset of the original example. Fig. 4.7 depicts the adder view for a part of the *jpeg* design space. This part is restricted to 2 or 3 load units. The bottleneck of the load units is avoided. The performance depends only on the number of adders, but is almost independent from the number of load units, either 2 or 3 units. The variation between minimum and maximum is very small. The designer can select the number of adders without worrying about the other unit types.

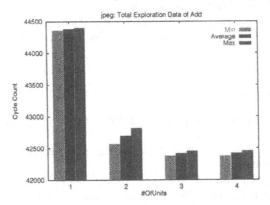

Figure 4.7.   Adder view when restricting the design space to 2 and 3 load units.

The average value shows how typical the bounds are. If the average is close to the minimum, this means that almost all processors achieve the minimum cycle time. There are only some processors where bottleneck constellation reduces the efficiency. On the other hand, if the average is close to the maximum, this indicates that there are only few processor constellation which achieve good performance. These are the processor structures that fit very well for the current application domain. The designer must spot these processors in the large number of processors that provide only modest performance.

Up to this point the discussion was focused on a single unit type. But an indicator is missing for the most significant unit type to look at. In the *jpeg* example, using a single adder is certainly a bottleneck. Even the highest cycle time with two adders is better than the best performance with one adder. But the results for 2 and 3 adders depend very much on the constellation of the other unit types. So the designer might select 2 adders and start the investigation of the load units. In fact this investigation is carried out on a subspace $\mathbf{DS'}$ of the original design space $\mathbf{DS}$

$$\mathbf{DS'} = \widetilde{\mathbf{DS}}(P(adder) = 2) \subset \mathbf{DS} .$$

This will change the remaining unit views since they must be recalculated for $\mathbf{DS'}$ now. Each selection of a specific unit number for a certain unit type reduces the design space to a subset of the original design space. It is important that these restriction starts with the most important unit type. The selection of the important unit type will influence all the other unit types. Therefore a summary view is required that points out this important unit type.

The summary view **View$_{DS}$** calculates a 6-tuple for every possible unit view, i.e.,

$$\textbf{View}_{\textbf{DS}} \; : \; \tilde{L} \to \mathcal{R}^6 \; , \quad \tilde{L} \subset L \; ,$$

where $\tilde{L}$ is a subset of the set of all unit types. It contains all those unit types where the exploration range is not completely restricted

$$\forall \, unit_i \in \tilde{L} \; \Longrightarrow \; |DS_i| > 1 \; .$$

The value of the summary view is defined as follows

$$\textbf{View}_{\textbf{DS}}(unit_i) = \begin{pmatrix} \underline{Min}(unit_i) & \overline{Min}(unit_i) \\ \underline{Avg}(unit_i) & \overline{Avg}(unit_i) \\ \underline{Max}(unit_i) & \overline{Max}(unit_i) \end{pmatrix} \; .$$

That is, two bounds are generated for every element from the unit view. $\overline{}$ denotes the upper bound and $\underline{}$ denotes the lower bound. For example, $\overline{Min}$ and $\underline{Min}$ are calculated as follows

$$\overline{Min}(unit_i) \quad \equiv \quad \max_{P(unit_i) \in DS_i} \{Min(unit_i, \textbf{DS}, P(unit_i))\} \; ,$$

$$\underline{Min}(unit_i) \quad \equiv \quad \min_{P(unit_i) \in DS_i} \{Min(unit_i, \textbf{DS}, P(unit_i))\} \; .$$

These are the bounds over all unit numbers in the exploration range. As an example, Fig. 4.8 shows the derivation of the minimum bounds from the adder view. The *mpeg_play* program with the *moglie.mpg* sequence as data input is depicted. The adder view contains exploration data for 1 to 5 adders. The derivation of the bounds, e.g., on the minimum, can be considered as projecting the complete unit view into a single 6-tuple of the summary view. In case of the minimum cycle time, there are five values in the unit view (the leftmost bars). The upper bound on the minimum is obtained for processors with just a single adder. The lower bound is reached for processors with five adders. Taken together, both bounds form the leftmost pair of bars in the summary view.

The variation of the bounds (e.g. $\overline{Min}$, $\underline{Min}$) shows how important a unit type is. If the bounds differ very much (e.g., the adder in Fig. 4.8), this means that changing the number for this unit type changes the performance very much. On the other hand, if these bounds are similar (e.g., the comparator in Fig. 4.8), there will be no performance difference when increasing the number of units for this particular type. The selection of suitable numbers of units should start with the unit type where the largest variation of the bounds is

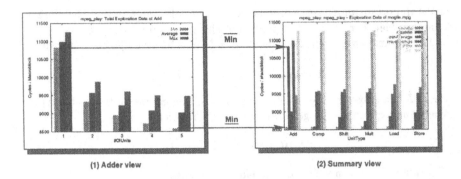

(1) Adder view   (2) Summary view

Figure 4.8. Derivation of the summary view from the different unit views. The complete unit view is projected into a 6-tuple of the summary view. The *mpeg_play* program and the *moglie.mpg* data sequence were explored.

encountered. This suggests the following procedure for analyzing the combined exploration data:

1. Let $\tilde{L} := L$ and $\mathbf{DS}' := \mathbf{DS}$, i.e., the analysis starts with all unit types and the complete design space.

2. The summary view $\mathbf{View}_{\mathbf{DS}'}(unit_i)$ is calculated for all units in $\tilde{L}$. The variation of the bounds is compared to determine the most important unit type $unit_{important} \in \tilde{L}$.

3. The unit view $\mathbf{View}_{unit_{important}, \mathbf{DS}}$ for the important unit type is calculated. A suitable number of units $P(unit_{important})$ is selected for this important unit type.

4. The design space is restricted to all processors which have exactly $P(unit_{important})$ units of the type $unit_{important}$

$$\mathbf{DS}' := \widehat{\mathbf{DS}}(P(unit_{important})) \,,$$

and the remaining unit types are considered, i.e.,

$$\tilde{L} := \tilde{L} - \{unit_{important}\} \,.$$

5. The procedure continues with step (2) until all unit types have been explored, i.e., $\tilde{L} = \emptyset$.

The procedure is a guideline for selecting the most suitable combination of functional units in the processor data path. It reduces the design space to one or a few processors.

Register File Exploration

The last exploration step is the register file exploration. It adapts the structure of the register file to the chosen processor structures. The result is a further reduction of the design space

$$\mathbf{DS_{register}} \subset \mathbf{DS_{combined}} \quad where \; |\mathbf{DS_{register}}| \ll |\mathbf{DS_{combined}}| \; .$$

Three parameters must be defined for the register file

$$\mathbf{RF} = (RF_{in}, \; RF_{out}, \; size) \; .$$

RF_{in} (RF_{out}) denotes the number of write (read) ports and $size$ denotes the number of words that can be stored in the register file. The size and the number of ports should be kept as small as possible to make the register file fast and to reduce the silicon area. The register file provides the input operands for the functional units and it stores the results of the computation. The register file is important for the utilization of the functional units. It is the source and the destination of most data. A very small register file would hinder the functional units. There are some other sources of operands: the load units, the immediates, and the forwarding of results from functional units. The immediates are taken directly from the instruction word of the processor. They are used for defining constants. The forwarding bypasses the instruction pipeline by directly forwarding the result of the execution stage as input to another functional unit. This avoids the latency of the write back and it reduces the traffic to the register file, i.e., a small register with less ports can be used. However, the forwarding is limited in its application because only adjacent operations in the data flow graph, which are scheduled in consecutive time slots, can profit from the forwarding.

This means both the immediates and the forwarding are limited to special cases of providing operands. The last alternative source of operands are the load / store units. They are the connection to the memory hierarchy. In principal, all operands might be stored in memory. But the memory is typically slower than the register file [76]. Even in case of a cache hit, the fetch from the memory might take 1 or 2 cycles. It is also necessary to calculate an address for every memory fetch. This increases the operation count. Therefore most modern processors and compilers try to store all intermediate results in the register file. The register file will be the dominant source of operands for the data path. The memory is mainly used for fetching the initial inputs of a computation and for storing the final results. The most suitable number of load / store units was already calculated in the combined exploration. For that calculation it was assumed that a large register file is used which stores as many intermediate values as necessary. This is the best condition where the lowest memory traffic occurs. But the register file would be prohibitively large.

The register exploration replaces such an oversized register file with a smaller file that does not reduce the performance significantly. The exploration assumes an unlimited forwarding since the interconnection network was not restricted in the design exploration. An estimate on the necessary number of ports for the register file can be obtained by looking at the flow equilibrium for the operands. All operands consumed by the functional units during a clock cycle must be provided by one of the data sources (memory, register file, immediates, forwarding) where the register file will dominate all other operand sources. The average number of transfers will be used as an estimate for the required number of ports in the register file. Let $(\Omega \times F \times T, \mathbf{B}, \mu)$ denote the probability space defined in section 4.2. A random function $Operands$ is defined

$$Operands \ : \ T \times \mathbf{DS} \rightarrow \mathcal{Z} \ .$$

The function denotes the number of input operands $Operands(type, \mathbf{P})$ required by the processor $\mathbf{P}$ for the operations of type $type$. Assuming a one-to-one correspondence between the operation types and the unit types, as above. The number of operands is given by

$$Operands(type, \mathbf{P}) = Operands(type) \cdot P(type) \ ,$$

where $Operands(type)$ is the number of inputs for the unit that executes the operations of type $type$. For example, a typical RISC addition will require two input operands. The operand produced by a memory fetch of the load unit is counted as a negative number. Let $E(Operands)$ denote the expectation value

$$E(Operands(\mathbf{P})) = \sum_{type \in T} InstMix(type) \cdot Operands(type, \mathbf{P}) \ .$$

In this case the estimate of the number of register read ports is given by

$$RF_{out} = \frac{1}{\alpha} \cdot E(Operands(\mathbf{P})) - forwarding - immediates \ ,$$

where $forwarding$ ($immediates$) denotes the average number of forwarding results (constants from the instruction word) that is used per cycle. α accounts for the throughput reduction that will occur when mapping the application on the processor. The equation states that the register file should have as many ports as operands are needed by the data path, subtracting operands which stem from immediates or forwarding. In a similar way the number of register write ports is given by

$$RF_{in} = \frac{1}{\alpha} \cdot E(Out(\mathbf{P})) - directForwarding \ .$$

$E(Out(\mathbf{P}))$ denotes the average number of results that is produced by the data path. $directForwarding$ denotes the average number of forwards that completely bypass the register file, i.e., the intermediate results are directly used in the next cycle and no further reference is made to them.

The estimates are used to specify suitable exploration ranges for the register file exploration. The register exploration varies the register parameters and determines the cycle time for every processor in the design space $\mathbf{DS_{register}}$. The code is scheduled for these processors and a complete register allocation is performed. A maximum forwarding is permitted. A typical example is depicted

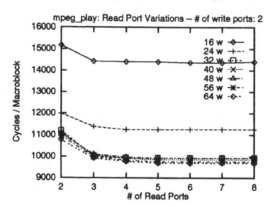

Figure 4.9.   A typical example of a register file exploration. The *mpeg_play* program and the *moglie.mpg* sequence were explored.

in Fig. 4.9. The data of the *mpeg_play* program and the *moglie.mpg* sequence are shown. The x-axis shows the different numbers of read ports that were explored. The register size is used as parameter of the various curves. The figure shows the results for register files with 2 write parts. The cycle times improve up to a register size of 32 words. Increasing beyond this size is not justified by the performance gains. Approximately 4 ports are useful for the register file. This slightly underestimates the number of ports because the forwarding was not limited and a complete interconnection network is assumed, i.e., all units can read results from all other units. Therefore the scheduler can perform more direct data transfers. Thus reducing the need to store intermediate results in the register file. A real processor has to restrict the forwarding to reduce the size of the interconnection network to reasonable values. This is addressed in the next section.

DATA TRANSFER ANALYSIS

The data transfer analysis provides the data for designing the InterConnection Network (ICN) of the processor. The interconnection network routes the data among the functional units of the processor. Data transfers to the memory are done only be the load / store units of the processor since a load-store architecture is assumed as the basic processor architecture paradigm. Two types of transfers can be distinguished: transfers to the register file and forwarding of results. The register file acts like a small temporary buffer. It allows to store intermediate results that will be used at some subsequent cycles. The register file is not efficient when results are consumed by operations which directly follow the producing operation in the instruction pipeline. In this case, the consuming operations must be delayed until the result is written back into the register file.

This inefficiency can be avoided, if the processor provides a forwarding of the result, i.e., the result is directly transferred from the producing unit to the consuming unit. Seen from the instruction pipeline, this means, the operation in the execution stage can directly use the result of the operation that was most recently executed. There is no need to postpone the operation until the producing operation has passed the write-back stage of the pipeline. Designing the interconnection network means to select the forwarding connections and it means to determine the connections between the functional units and the ports of the register file. A trade-off is performed which is explained by looking at the extreme cases of the interconnection network design.

A complete interconnection network would provide the best possibilities for routing the operands. In this case there are busses for all operand sources (immediates, register read ports, unit outputs). The inputs of all functional units are connected to all these busses. The interconnection network provides a maximum of forwarding since all results can be directly transferred to all possible units. Furthermore, all register ports and all immediates can be used by all functional units. This means, there will be no resource conflicts because of a missing bus and all possible transfers can be done without restrictions. Therefore, no idle cycles occur due to the interconnection network. Unfortunately, this interconnection network would be too complex and too slow to be useful in praxis. A large silicon area will be occupied by the busses and each of the input ports needs large multiplexers to connect to all these busses. Furthermore, the bus drivers will face a high load which makes them slow and increases the power consumption of the chip. In summary, the complete interconnection network would be useful from the architectural point of view, but all the advantages are negated by the practical limitations set in the implementation technology.

The second extreme is a very small interconnection network. Only a minimum fan-in is used for all unit inputs. No forwarding is applied. In this case, the area is small and the load of the busses is minimal. Only small multiplexers are required at the inputs of the functional units. A high clock frequency can be achieved, but there will be a large number of idle cycles where units have to wait for input operands. The utilization of the functional units is low because the interconnection network is a bottleneck.

In between these two extremes is a well designed interconnection network. The fan-in of each input is kept low without sacrificing to much performance. The data transfer analysis is used to find this point where the architectural advantages of the complete interconnection network and the implementation advantages of the minimal interconnection network are well balanced. The data transfer analysis is subdivided into two steps:

Analysis of the transfer probabilities. This step is similar to the instruction mix analysis. It investigates the probability that a transfer between different operation types occurs. These data are used to tune the forwarding connections. If a transfer occurs only with low probability, the performance will not decrease very much, if this transfer is done via the register file. This can be exploited to reduce the number of connections without sacrificing much performance. On the other hand, if a transfer occurs frequently, this is a good candidate to be supported by a direct forwarding. The result can be directly transferred from the producing unit to the consuming unit.

Bus exploration. This step is comparable to the design exploration of the functional units. It determines the cycle time for processors with different numbers of busses. A variation of a complete interconnection network is used for this investigation. All input ports are connected to all busses and the busses can be used to transport all results. But the total number of busses is restricted. In this case the type of transfer is not considered but the number of transfers that might be done in parallel. Providing just a small number of busses will produce frequent resource conflicts for these busses. The functional units will idle because of missing input operands. During the exploration, the number of busses is gradually increased. The performances asymptotically approaches the performance of the complete interconnection network. The generated data are used to determine the point where the performance is nearly as good as for the complete interconnection network but the number of busses is as small as possible.

Transfer Probabilities

The explanation starts with the analysis of the data transfer probabilities. Let

$$\tau_{i,j} \equiv (type_i, type_j) \ , \quad type_i, type_j \in T^2$$

denote the data transfer from an operation of $type_i$ to an operation of $type_j$, i.e., the operation of $type_i$ produces a result which is consumed by an operation of $type_j$. The total number of these transfers for a data block is denoted by $\frac{trans_\omega}{block}$. ω is the data sequence that was used as input to the application program. The analysis of the transfers per block can be compared to the $\frac{op}{block}$ investigation in the requirement analysis. In fact both values are closely related when considering typical RISC operation. Normally, each operation will consume one or two input operands and produce one result. Improving the operation count will also tend to improve the transfer count. In most cases the designer can use just the operation count when improving the algorithms or the software implementation (as described in section 4.2). The transfer count is only used for designing the interconnection network. Therefore the measurement is refined w.r.t. the different types of transfers. A probability space $(\Omega \times T^2, \mathbf{B}, \mu)$ is used. $\Omega \times T^2$ denotes the set of all sequences and all transfers that might happen. $\mathbf{B}$ is the σ-field of all subsets of events that can be formed and μ denotes the occurrence probability of these events. The probability of using a certain transfer type in a specific sequence is denoted by

$$DT_\omega : T^2 \ \rightarrow \ [0,1] \ ,$$

where $DT_\omega(\tau) \equiv \mu(\tau|\omega)$. The data sequence index will be dropped when the specific sequence is not important for the discussion. The transfers per block of a specific type are given by

$$\frac{trans(\tau)}{block} = DT(\tau) \cdot \frac{trans}{block} \ .$$

This value is calculated by examining the transfers that occur during the programs execution. Realizing this transfer in hardware means to switch a connection from the unit result bus to a unit input port. Let CPC denote the maximum number of connections that can be switched by the processor in each clock cycle. Again a one-to-one relationship is assumed between the unit types and the operation types. The design goal of the interconnection network is to use as few connections as possible. In the ideal case, each connection would be used for transferring operands in each cycle. Let *connUtil* denote the fraction of the maximum transfer capability that is actually used in the application. The design equation for the interconnection network is defined in a similar way as equation 4.3

$$CPC(\tau) \;=\; t \cdot \frac{DT(\tau)}{connUtil(\tau)} \;,$$

$$where \; t \;\equiv\; \frac{trans}{block} \cdot \frac{blockRate}{f_{Clk}} \;.$$

t denotes the required transfer rate of the application program. Up to this point the analysis is almost identical to the requirement analysis. This suggests to use the same procedure, that was used to determine the functional units, when selecting the forwarding connections. However, there is an important difference: the larger number of possible transfer types. The number of operation types is only $|T|$ while the number of transfer types is equal to $|T|^2$. This in turn implies that the likelihood of using a specific transfer type can be very low for a large subset of the transfer types. Assuming for example a uniform probability distribution for both the operation types and the transfer types, means that an operation is executed with probability $1/|T|$ whereas a particular transfer is executed only with probability $1/|T|^2$. Providing dedicated connections for all possible transfer types would waste hardware resources due to the low utilization of these connections. The situation is aggravated, if the transfers are refined into the transfers to particular inputs of an operation. Often, the inputs are not commutative, e.g., in case of a shifter. This would increase the number of different transfer types and thus lower the probability of using a particular type.

The situation will be examined in more detail by looking at an ideal forwarding situation similar to the ideal scheduling situation that was used in the requirement analysis. We start the thought experiment by looking at an imaginary processor (Fig. 4.10). The processor contains the number and type of functional units as determined by the design exploration, but the interconnection network is special. It allows an unlimited forwarding except for the resource constraints on the number of available connections. This is achieved by to components: a shuffle network and result delays. The shuffle network allows to exchange the busses that are used to transfer results from units of the same type. For example in Fig. 4.10, there are two result busses for transporting adder results. The shuffle network allows to exchange results coming from the two adders. Due to the shuffling network, the binding of an addition to one of the two adders needs not to be considered. Next, the delay circuits allow to produce arbitrary delays from Z_+ for all the results. This means, the results can be reordered by the delay circuits, but the allocation to the

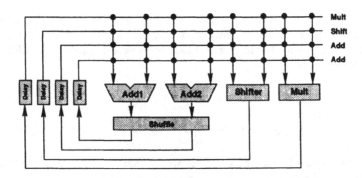

● : multiplexer connection to an input port

Figure 4.10.   Imaginary processor for the ideal forwarding situation.

busses is not changed. These delay circuits act like imaginary register files of infinite size which are devoted to results of a particular operation type. First of all, we assume a complete interconnection network as depicted in Fig. 4.10. In addition, an ideal scheduling is assumed, where even operations, that are scheduled in the same cycle, can exchange results. Hence the ideal forwarding processor would correspond to the processor structure used at the beginning of section 4.3.

In addition, we assume RISC like operations where each unit produces one result and consumes two input operands. The number of result busses in this case is given by $\|\mathbf{P}\|$. The number of active units is reduced by the ideal throughput reduction. Therefore the following number of results is produced in each cycle,

$$\frac{results}{cycle} = \frac{1}{\alpha} \cdot \|\mathbf{P}\| \ .$$

Since each of these operations needs two input operands, this results in a total number of input operands of

$$\frac{inputs}{cycle} = \frac{2}{\alpha} \cdot \|\mathbf{P}\| \ .$$

The number of transfers is the maximum number of results or inputs that must be transported in each cycle.

$$t_{ideal} = \max\{\frac{inputs}{cycle}, \frac{results}{cycle}\} \ .$$

It is assumed that all operands are delayed versions of the results, i.e., all operands can be provided by the delay circuits. The maximum number of transfers in case of RISC like operations is therefore

$$t_{ideal} = \frac{2 \cdot \|\mathbf{P}\|}{\alpha}$$

The circuit of Fig. 4.10 contains a crossbar with

$$CPC = 2\|\mathbf{P}\| \times \|\mathbf{P}\| = 2 \cdot \|\mathbf{P}\|^2$$

connections. There is a connection from each possible result bus to each input port. The total utilization of the connection network is therefore limited by

$$connUtil \leq \frac{t_{ideal}}{CPC} = \frac{1}{\alpha\|\mathbf{P}\|} \ .$$

For each unit port there is at most one active connection, but the interconnection network provides $\|\mathbf{P}\|$ connections. The utilization for each transfer type is therefore bounded by

$$\forall \, \tau \in T^2 \ : \ connUtil(\tau) \leq \frac{1}{\|\mathbf{P}\|} \ .$$

In the example of Fig. 4.10, only 25% of the connections would be used in every cycle. This is already the ideal case for the complete interconnection network since we have assumed arbitrary delays for the results so that restrictions on the register size were excluded. The utilization of the complete interconnection network gets worse if the number of functional units in the processor increases, i.e., if $\|\mathbf{P}\|$ becomes larger. As a consequence, the performance of a large processor will be reduced not only by the available instruction-level parallelism but also by the large number of connections which in turn will reduce the clock speed. This means a large processor with a complete interconnection network will have poor performance from the architectural *and* from the implementation point of view.

This poor performance is caused by connections which the hardware provides for transfers that seldom occur in the application. Dedicated connections should only be used for transfers that occur frequently. All other transfers must be implemented in a different way. First, the important transfers are selected from the set of all possible transfers. A forwarding threshold

$$\delta_{fwd} \in \mathcal{Z}_+$$

is defined. The threshold denotes the number of transfers that must occur per cycle to justify the support by a direct forwarding. The set of forwarding connections FW is given by

$$FW \equiv \{\tau \in T^2 \; : \; t \cdot DT(\tau) > \delta_{fwd}\} \; .$$

It contains all the transfers where the number of transfers is larger than the forwarding threshold. Transfers in this set occur with a probability

$$DT(FW) > \frac{|FW| \cdot \delta_{fwd}}{t} \; .$$

Next, an alternative realization is necessary for all the other transfer types, i.e., the transfers in $T^2 - FW$. These transfers occur not frequently enough to justify the dedicated connection, but they are necessary for the correct processing of the application. These transfers are realized by the multi-ported register file of the VLIW processor in Fig. 2.6. In fact, the multi-ported register file can be considered as a combination of a shuffle network for several different unit types and a delay circuit. It is not an ideal delay circuit as in Fig. 4.10 because its size is limited. Therefore the number of words which can be stored at the same time is limited as well. Additional storage can be provided by spilling data into the memory. But this causes a severe performance penalty. The register file exploration of the preceding section has measured register file size where the spilling is minimal. An estimate of the required number of ports was determined as well. This estimate can be refined by considering the forwarding connections that are used when implementing only the most important forwarding connections. In this case the number of register write ports is approximately equal to the number of results produced in each cycle minus the number of implemented forwarding connections

$$RF_{in} \approx \frac{1}{\alpha} \cdot \|\mathbf{P}\| - |FW| \cdot \delta_{fwd} \; .$$

The connection of the unit result ports to the register write ports is guided by two conditions. First, the units might be grouped in such a way that each group produces the same number of results per cycle. This grouping is done according to the results of the instruction mix measurement. The second condition is the total number of connections to a single register port. This number must be small to avoid large multiplexers which would slow down the clock speed of the processors. The connection of the register read ports and the input ports of the functional units is done by considering the data transfer probabilities to a certain operation type and by reducing the fan-in of each unit input port as much as possible.

Bus Exploration

The data can be used as a guideline for the design of the interconnection network in the cosynthesis. In summary, the discussion has focused on the types of transfer. A set of the most important transfer types was determined. These are the candidates for the implementation of the forwarding connections. The remaining transfers are realized via the register file. The connection between the units and the register file ports is guided by the data transfer statistics and the instruction mixes. The connections should distribute the traffic to the register ports as evenly as possible. This is achieved by grouping transfers so that they have equal probability. All these investigations are based on the dynamic transfer type probabilities. The scheduling of these transfers was not considered. It was implicitly assumed that operations which are adjacent in the data flow graph can be scheduled in adjacent time slots. This is not always the case since there might be resource conflicts on the functional units or other input operands might be missing so that the operations cannot be scheduled in adjacent time slots. In this case the forwarding connection cannot be used and the result must be stored in the register file for use in a later time step. This means the forwarding efficiency is reduced at the expense of an increased register file traffic. As a consequence, the schedule is increased and the average number of transfers per cycle is reduced. Therefore, it is important to know the average number of transfers that will occur *after* scheduling the programs. These data are determined by the bus exploration. The processor structure, determined in the design exploration, is used as the basis of the bus exploration. It assumes no restriction on the number of forwarding connections but a restriction on the total number of busses. This differs from the design exploration. In the design exploration it was assumed that there is a large number of busses. So whenever a bus is necessary to transport a result, there is a free bus available. This condition is changed in the bus exploration. Now, the number of busses is limited. Resource conflicts occur for the busses. It is determined, at which point the performance suffers from reducing the number of busses. How much "bus shortage" is tolerable? Let b denote the number of busses in the processors. A processor with b busses can transfer at most b data per cycle, i.e.,

$$t \leq b .$$

Again we can consider two extreme cases for a processor with RISC like operations. Using a complete interconnection network will give the cycle time that was already determined in the design exploration $\frac{cycle(\mathbf{P})}{block}$. The other extreme is a processors with just two busses $b = 2$. In this case at most one operation can be done in each cycle. Therefore the cycle time

$$\frac{cycle(\mathbf{P}, 2)}{block} \geq \frac{cycle(\mathbf{P}, b)}{block} \geq \frac{op}{block}$$

is at least as large as the number of operation per block. It might be larger if operations are pipelined and depended operations must wait until the operation has reached the end of the pipeline. The pipeline latency of the units is added as idle cycles. This was discussed above.

The missing element are the data that fill the equations. The next chapter describes the design tools for measuring them.

5 DESIGN TOOLS

So far we have looked at the design project and sketched the design methodology. The preceding section has investigated the analysis framework for deriving the principal processor structure. The approach is based on a careful analysis of measured data. It allows to evaluate trade-offs at a very abstract level, without spending time on implementation details. However, the analysis approach needs measured data as the basic input. In embedded system design, these data are specific to the current design. Tools are necessary to measure them. This section describes these design tools in detail. Before doing so, however, we will discuss two different types of tools: all-automatic tools and designer centered tools. The discussion will show that there is a shift of emphasis. In particular, the preparation of data for the designer becomes more important in the latter case. The designer is only able to influence the design, if the amount of data is not extremely large. The data must be prepared so that the resulting parts are manageable. This raises a couple of questions. What kind of data visualization is the most powerful? How can the designer spot the interesting points of the design? Which is an efficient way to access the data?

All-automatic tools take an abstract specification and map it into a more detailed specification with little or no user interference. An example is a synthesis tool which reads a VHDL specification on RT level and generates a gate-level specification. Another example is a normal compiler which takes a program written in a high-level language and transforms it into an assembler program. In the ideal case these tools do not rely on the user to control the mapping process. Therefore it is not important to inform the designer about the internal decisions of the mapping process. In contrast, the designer centered tools rely on the designer's control of the mapping process. The designer has to provide knowledge. Often, the tools will perform only subtasks of the complete design process. Most of the knowledge based decisions are done by the designer. Therefore the visualization and access to the internal information is a crucial aspect. Otherwise, the designer cannot control the design process. Typically, the data format used by the tools is not directly useful for the designer. The designer centered tools must spend more effort in preparing data for the designer. The analysis of fine-grained design data is one of the most important aspects for designer centered tools.

The main goal of design automation (especially in competitive market segments like multimedia) is the reduction of the design time. This in turn reduces the time-to-market which gives higher revenues. Therefore, an all-automatic approach seems favorable because it would require no human design time (at least in the ideal case) and thus reduce the costly manual design time to zero. The major disadvantage of an all-automatic approach is the missing designer knowledge (see chapter 4). This part gets more and more important when the level of abstraction is raised. On lower levels of abstraction it is often possible to use a single paradigm for all kinds of different designs. The design tools are adapted to this paradigm. An example is a standard cell library that is used in logic design. The optimization is done within this framework. On higher levels of abstraction, it is crucial to integrate knowledge from many different domains. Often, the major design improvements are obtained when the designer deviates from the standard approaches. It is difficult (if not impossible) to include this kind of reasoning into automatic tools. Artificial intelligence has striven to build up knowledge bases [123] for many years, but it is still not possible to build up knowledge bases for a large spectrum of different fields. Exactly this interdisciplinary knowledge is the crucial element in system-level design. Only the successful combination of ideas from many different fields can make the design of the complete system successful. Currently, only the designers can provide this knowledge. Therefore, the designer centered approach exploits this knowledge by informing the designer about the possible trade-offs. The final decision is left to the designer. The tools move from optimization systems into information management. In this way the machines ability to speed up monotonous tasks is combined with the designers global knowledge about the system design. The designer will use the tools to solve certain sub-problems, for example, trying out a large number of different processor structures in the design exploration. Often these problems cannot be solved analytically when dealing with complex systems. Therefore the design tools are essential as instruments for solving numerical sub-problems. The resulting data are the basis of the design decisions. It is even possible to generate measured data for processors which are not actually realizable. An example is the relaxed processor which is used in the design exploration when calculating performance bounds. It is not possible to determine these bounds without the help of the design automation tools.

But this close interaction of the design tools and the human designer requires a different type of tools. The results of the sub-problems must be prepared for the human designer. Typically, this will involve a visualization of the data. It is important, that the visualization scales up to complex designs. For example, a gate-level schematic is an useful aid for displaying a small circuit. But if it is used for very large processor circuits, the result is too complex to provide useful information for the designer. The designer is overwhelmed by details. The tools provide too much information in an unstructured way. A hierarchical approach, which abstracts major parts of the circuit, is necessary. In essence, the analysis part which prepares the data for the designer acts as an information filter that extracts the most important informations from the large set of design data. Producing the data without preparing them is useless. The following sections will describe the structure of tools for both, the quantitative analysis and the cosynthesis, in more detail. The main focus is on the individual tools. The configuration of the tools into design flows and the integration of these tools into a networked computer environment will be addressed in chapter 6. A detailed description of the design tools is given in [198].

ANALYSIS TOOLS

The quantitative analysis gradually defines the structure of the embedded system. It gives a picture of the system performance and it provides information about the performance of possible alternatives. The general methodology was derived in chapter 4. The calculations frequently relied on measured analysis data. The generic structure of the analysis tools is depicted in Fig. 5.1. Two major parts can be distinguished: the *measurement* and the *analysis* part.

The measurement part is typically the most time-consuming task. It performs a detailed investigation of the software (in case of the requirement analysis) or it generates a large number of alternative processors and checks the performance of these alternatives (in case of the design exploration). The inputs are the program(s), that should be investigated, and an ensemble of typical data sequences. As described in chapter 3, C/C++ programs are used as input

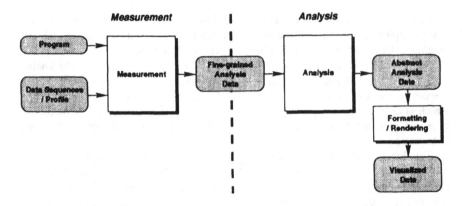

Figure 5.1.   Principal structure of the analysis tools.

for designing the compression system. The main advantage of using a standard language like C/C++ is the reuse of existing software. Typically, a large number of programs exist for virtually all types of applications. Furthermore, there is a well established programming environment. It gives a significant advantage for fast software development.

The data sequences are selected from the set of all possible sequences in such a way that they form a representative ensemble of inputs. In case of the analysis for the video compression system, a large number of sequences exists which are often used for analyzing the performance of the coding algorithms. The analysis in this thesis uses these standard sequences to make the results comparable with the algorithmic investigations performed in other studies. The measurement determines the design data for all these input samples. This is a computationally intensive task. The results are used to extrapolate the performance distribution for the set of all input sequences. This in turn can be used to check the soft real-time condition of the embedded system. A full simulation of the video data sequence is often extremely time-consuming. Therefore it is sometimes (e.g., in the design exploration) replaced by using profiling information from previous simulation runs. The profiles contain the execution frequencies of all basic blocks in the program and the transfer counts between adjacent basic blocks. In most cases these data are sufficient to reconstruct the execution behavior when processing a certain data sequence.

The measurement of the program can be done on different levels of abstraction. For example, the C programs can be instrumented and analyzed. The main advantage is the fast execution time and the machine independence. A disadvantage is the reduced accuracy in comparison with investigations at the instruction level. The analysis for the video compression system is performed on the instruction-level. The program is translated into a sequential assembler program. The sequential program defines the execution order for a simple scalar processor that executes only one operation at a time and does not use any pipelining [179]. The types of operations that are allowed in the assembler program are defined in the instruction-set. The instruction-set is the "contract" between the compiler and the processor hardware [152]. The hardware must ensure that every operation in the instruction-set can be implemented by the hardware. The compiler ensures that the semantics of the high-level language are represented by language from the specified instruction set. The MOVE [80] instruction set is used in this thesis. It is a typical RISC instruction-set, i.e., the operations can be directly implemented by hardware units. In this way the compiler has a tight control over the hardware operations. This is especially important in VLIW processors where the compiler performs the scheduling. While the analysis in this thesis focuses on the assembler level input, the overall methodology holds for higher-level analysis as well.

The dynamic behavior of the assembler program is measured for different input sequences. The results are fine-grained analysis data. For example, the execution profile of an instruction-set simulation, or the processing times for a large number of different processors. These data reflect the operation of the system in detail. However, the data are typically too detailed to be of direct use for the designer. The manual examination of a large number of processing times is very time consuming, if done by hand. Therefore the analysis tools are necessary to transform the fine-grained data into a more abstract representation which is useful for the designer. Often this part is interactive because the designer has to decide on certain parameters of the analysis. For example, when selecting the most important functions, the designer has to decide on an appropriate value of $n_{important}$, or when displaying the summary view of a design exploration, the designer must select the most important unit view, etc. Hence, it is important that the analysis part is kept very fast to allow the interactive operation without significant delays for the designer. This is already taken into account when generating the fine-grained data: as much calculation as possible is pushed into the measurement phase.

An alternative to the complete interactive operation is the hyperlink framework presented in chapter 6. In this case the analysis prepares automatically all analysis data which are *typically* used. The user can navigate among these data with a conventional hypertext browser. This means all the calculation is done before hand and the designer can use the results without any waiting time. This is especially useful for performing complex analysis tasks where even the interactive preparation of the analysis data would be very time-consuming. It is advantageous to generate two versions of the abstract data: a computer readable version and a formatted version for displaying it to the user. A convenient format for the computer readable format are TAB separated ASCII files. This is an exchange format which facilitates the processing by other UNIX programs [153] and it can be used as input to many text processing systems or spreadsheet programs. It supports further processing of the abstract analysis data, e.g., to combine abstract data from different analysis runs. However, the ASCII format would not be convenient for displaying it to the user. Formatting and rendering tools are necessary which take the TAB separate table as input and generate bar charts and graphical views of the data. Retaining both versions of the analysis data provides a maximum flexibility. The data are visualized for the designer and future reuse by other programs is possible.

The preparation of the abstract data fulfills two tasks. First of all, the fine-grained data are combined in different ways to make it easier for the designer to spot the important information within the large number of analysis data. An example are the different views used in the design exploration. In this case it would be difficult for the designer to work directly with the multi-dimensional data. The tools prepare the views described in section 4.3. They support the designer in spotting the most important unit types without looking at the complete set of exploration data. The second task is the normalization and preparation of the data for the calculations in the design process. For example, most of the analysis in chapter 4 was carried out on a data block level. This allows to scale results to different frame formats and different frame rates. It would be very time-consuming for the designer, if he/she must perform the normalization manually. Whereas it is easy to perform this normalization before preparing the abstract data tables. Unfortunately, operations like normalization of result data are very application dependent, e.g., the normalization of the processing times can only be used if the processing times are linear with the number of data blocks. Therefore the final preparation of the abstract analysis data must be controlled by programs which might be adapted to different application domains.

COSYNTHESIS TOOLS

This section describes the tool structure of the cosynthesis environment. A generalized cosynthesis is considered. The generation of the hardware description and the implementation of the software is addressed. The cosynthesis is not restricted to an architecture template of a general-purpose processor and dedicated coprocessor as in [67] [47]. For the video compression system, a processor hardware description of a VLIW processor is generated and a compiler back-end for that particular processor. The compiler back-end is used to translate the sequential assembler code into processor dependent parallel code. The sequential code is produced by the compiler front-end. In other words, the generated compiler produces the software for the generated hardware description. The cosynthesis tools derive both parts of the system, the hardware and the software, from a single description. This means that the consistency of both descriptions is automatically secured. The feature is especially important when different processors are designed for evaluating a number of design alternatives. The designer can use a single C program for all different processors and the opcode can be adapted by using machine specific compiler back-ends which are produced in the cosynthesis.

The cosynthesis design process can be divided into three main steps:

1. An architecture framework is selected.

2. The parameters of the architecture framework are defined to adapt it to the application domain of the embedded system.

3. The design tools generate a hardware description on the RT level. The description is written in VHDL. A compiler is produced to translate the software.

The first step is the architecture selection. The video compression system uses an application specific VLIW processor as described in section 2.4. The framework provides a number of different parameters which are selected by the designer. The parameters allow to adapt the framework to different application domains without specifying too many details. This means the designer has two different levels of customization: (1) the selection of the architecture framework (e.g., a VLIW framework is suitable for high performance applications while a micro-controller framework would be more useful for embedded controller tasks) and (2) the selection of the parameters within the framework. The first level of customization is very general. Typically, a design group will use the same type of architecture for many different designs. The second level of customization is the real design level. This is the level, where the tailoring of the hardware and software to a particular design context is performed. The basis for this parameter selection is the quantitative analysis. The number of parameters defines the degree of automation. If there are only very few parameters, it means that most of the design is done automatically. As a disadvantage, the designer would have not much possibility to influence the design, i.e., he/she will spend not much time on the parameter selection but there is also no way to exploit the designers knowledge for improving the design quality. The other extreme is an architecture with very many different parameters. For example, a gate-level standard cell design might be considered as an architecture framework where the designer can even influence the types of gates and the interconnects between these gates. For complex designs this leads to extremely high design times because the designer must specify millions of gate instances and the connection among these gates. In between these two extremes are the interesting architecture frameworks where the number of parameters is sufficiently low to allow a fast design process, while giving the designer enough freedom to exploit his/her knowledge about the application for improving the quality of the design. The selection of the parameters is typically done by using a graphical tool (i.e., entering the block diagram) or by using compact description files.

The last step of the cosynthesis is the translation of this parameter specification into the final hardware and software description. A single parameter specification is used to let the tools control the consistency among the hardware and software domain. The architecture framework is necessary to provide a basis for the automation. The tools "know" the framework. Hence, procedures can be programmed to automate tasks within the framework. Again the generality of the framework defines how much automation can be achieved. If the framework is very generic, this means that the automation is extremely difficult because many different situations must be considered. If the framework is too specific, this limits its applicability to narrow application domains. The potential customer base for the design tools decreases. The tool structure for the cosynthesis environment is depicted in Fig. 5.2.

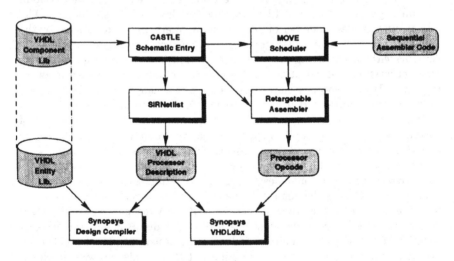

Figure 5.2.   Structure of the cosynthesis environment.

Schematic Entry

The central component is the schematic entry. It is used to define the parameters of the architecture framework. The VLIW framework of section 2.4 allows, for example, the selection of the number and type of functional units, the register file sizes, the interconnection network, etc. The schematic entry is used to define these parameters on a block diagram level. For example, the designer can select the units of the processor data path from a component library. Complex units like the register file have some parameters (e.g., size, number of read and write ports) that the designer can specify. For all other unit types only the instance name must be determined. After selecting the units of the processor data path, the designer will insert the connections among the data ports of the functional units. Only the main busses and the forwarding connections are specified by the designer. All inferior signals like the controlling signals, reset, and clock are inserted automatically when generating the hardware description. The schematic forms the basis for generating the hardware description and the compiler back-end. In addition to the schematic entry there is also a shell based interface. This is useful for specifying the processor (or parts of it) by other tools.

The hardware description is generated in two steps. First, all parameter dependent expressions are translated into constants. For example, the widths of the control signal for an address port of the register file depends on the size of the register file. The width can only be calculated after the parameter was specified by the designer. Once the netlist with fixed signal width is created, it is translated into a *SIRNetlist* [199]. The SIRNetlist is generic. It provides a number of output methods. These output methods allow to dump the netlist in different hardware description languages (e.g., VHDL, Blif). It makes the specification in the schematic entry independent of the design tools that are used for the final hardware synthesis. All output formats of the SIRNetlist can be used to connect design tools to the cosynthesis tools. New output languages can be generated by adding appropriate output methods for the SIRNetlist.

The compiler back-end consists of two parts: the MOVE scheduler and the retargetable assembler. The MOVE scheduler translates the sequential assembler code into parallel assembler code for the processor that was specified in the schematic entry. The schematic entry supports this by generating the appropriate machine description for the scheduler. The parallel code contains the timing information (i.e., all operations which are combined in one instruction are written on one line of the parallel assembler code) and the allocation / binding information. This defines the unit that will execute the operation. The retargetable assembler translates this information into the specific instruction format of the specified processor (Fig. 2.6). In particular, the encoding of the instructions is done in this step. The encoding depends on three elements: the allocation information generated by the scheduler, the machine description, and the component library. For example, the encoding of an addition will be different, if the addition is assigned to an adder or to an ALU. The retargetable assembler encodes each of the operations according to the unit type to which the operation was allocated. Finally, the operations of all the units are concatenated to form the final instruction word of the processor.

The backbone of the cosynthesis is the component library. From the hardware point of view it is an abstract library that defines processor components on a behavioral level. The library is written in VHDL. This allows to use the library components in conjunction with any commercial VHDL simulator or design compiler. The components of the library are generic. They use parameters for customizing the components. In this way, each of the components in the library describes a class of processor components. For example, the register file allows to specify the number of input and output ports as well as the number of words and the word width. This means the single register file of the component library describes all possible multi-port register file structures. It makes the library very compact without sacrificing its universality. The library is split into two parts: the interface declaration in the *VHDL-component* part and the internal behavior of the components in the *VHDL-entity* part. The latter is only used by the synthesis tool or the VHDL simulator. The cosynthesis tools treat all components as black boxes. The internal behavior is not used. Therefore the decision on the specific hardware realization is postponed until the final hardware design stage. This means the specification in the schematic entry needs not to be changed, if the designer decides on a full-custom realization of the component, instead of synthesizing it. The only component information that is used during the cosynthesis is the types of operation that are performed by the component and the encoding of these operations. It is necessary for generating the compiler back-end. The information is included as comments into the component library. The compiler-information extends the hardware description of the components into the software domain, i.e., a software description is done in addition to the VHDL hardware description. The cosynthesis tools are described in more detail in [198].

6 HTML-BASED CODESIGN FRAMEWORK

A short look back on the preceding chapters will summarize the main points that were addressed so far. A designer centered codesign approach was presented. It can be subdivided into two major parts: the quantitative analysis and the cosynthesis. The preceding chapter has outlined a number of different tools to support the various sub-tasks that have to be performed in this context. For example, the quantitative analysis can be subdivided into three major steps: the requirement analysis, the design exploration, and the data transfer analysis. Each of these steps can be further refined into a measurement part and an analysis part. The measurement consists of tools which perform the time-consuming fine-grained investigations of the system. The analysis part contains the tools that perform the information filtering. It generates the abstract data and visualizes them.

Imagining a designer using all the tools of the proposed methodology raises immediately a couple of questions. First of all, invoking a large number of different design tools by hand would be extremely time consuming and error-prone. Therefore it is necessary to automate the invocation of the tools, at least for the major design steps. The second problem is the repetition of similar design tasks with slightly different design context. For example, the quantitative anal ysis must be carried out for a number of different video compression programs and the analysis must be repeated for several input sequences. This means it is necessary to use templates for setting up the analysis. These templates can be reused for the next setup (maybe with some modifications to adapt them to the new context). In addition to automating the generation of the analysis data, it is also important to organize the results. In particular since an automatic quantitative analysis can generate lots of data in comparatively short time. Storing them as simple result files is not enough because the designer would be overwhelmed by this flood of information. It is necessary to organize them for a fast access of the important information. The context setup, that was used for the preparation, must be documented. Finally, a groupware context should be supported. In most cases the design of a complex embedded system will not be done by a single person, rather, the design will involve a group of people that have to cooperate on the project. In particular, designers with different background knowledge have to cooperate in embedded system designs. This is important because of the multi-disciplinary knowledge which is a crucial point for a successful design. This can even lead to design groups that are located at different sites. The design system should support this by providing seamless network integration.

This chapter presents an HTML (Hyper Text Markup Language) [190] based framework that offers a solution for all these problems, i.e., the way of *"putting it all together"* is addressed which makes things work. HTML is the language of the WWW (World-Wide Web) [19]. It contains commands for formatting text into headlines, lists, paragraphs, etc. Inline images con be included into the pages. The image files are stored as separate files (they are included by reference). Browsers often support the inlining of pictures in bitmap, GIF, and JPEG format. While all these are fairly conventional features of text processing languages, there are two important additional features: the links and the URLs (Uniform Resource Locator). The links are pointers to other pages, or other services. This allows to connect pages to other pages or to points within a page. An almost arbitrary cross-referencing is supported. The target of a link is specified by its URL (its position in a global computer network). A typical URL consists of the service type, the server address, the directory path (as seen from the root directory of the server), the filename, and the section within the file. This provides a simple but ingenious way of addressing files across the world. The displaying of the file, addressed by a link, depends on the application type of the file. The application type might be an HTML file or an MPEG video, etc. This concept extends the links into multimedia informations. All these features have contributed to the popularity of the WWW and the HTML language, making them globally accepted standards for exchanging multimedia information. This in turn has spawned many software developments for the

WWW. The software can be used directly in conjunction with the framework outlined in the sequel.

The framework can be used at three different levels of increasing sophistication. The necessary tool support is increased, if more advanced features of the framework are used. The following leve᷈ can be distinguished:

1. *HTML pages for organizing the design data.*

   This requires only minimal support from the design tool. HTML pages must be generated in addition to the rendered result files. The HTML pages are used to glue together the various result files and they define the setup that was used for generating the data. For example, a design exploration can store the ranges that were explored, the programs and data sequences that were investigated and so on. The main advantages of this basic HTML usage are the documentation of the results. Since the context is stored in addition to the raw result files. The pages can be exported via a normal WWW server. It supports the distribution of the data in larger design groups. The data can be accessed by a conventional HTML browser [189] like Netscape or Mosaic. No design tools are necessary to access the data, once the measurement has been performed. This is especially important for group members from a different design discipline. As an example, a hardware engineer can access the results of a software developer without using any special tools. The designers can concentrate on the tools which they use in their daily work. There is no need to learn the usage of all tools just to get informed about the latest results that were obtained by some co-workers of the project. Another advantage is the linking of the data. This provides a convenient way to represent the complex cross-relationships among the various components of the design. For example, the hardware design decisions can influence the selection of the algorithms or the software design. All this can directly be represented by links which connect the most suitable combination of design data. Furthermore, the design data can be referenced in many different contexts without changing their physical location. This is important if the same result files are used in conjunction with different investigations. For example, in a comparison of different programs and in a comparison of different data sequences. This also facilitates the extension of the design database for future designs. In this case parts of the design data can be re-used by the next design.

2. *Starting of design tools from the HTML pages.*

   In this step the design tools are used as link targets. Clicking on that link will start the tool. The tool is automatically started in the context that is defined in the HTML page. For example, the correct data are loaded into the tool. The designer can use the HTML pages as a graphical interface that can be modified by the designer himself/herself. This means the designer can configure the tools into design flows for an application by writing or generating appropriate HTML pages. Not only design tools can be used as call backs of the links in the HTML pages, but complete design processes can be started in this way. In that case, the call back defines a Makefile target which performs some updating operation (see section 6). This is convenient because the designer needs not to know the underlying directory structure when starting a design action. Only the HTML pages are used for the navigation on the design data.

3. *Integration of the complete framework.*

This step combines: (1) the editing of the HTML pages, (2) the modifications of the file structure, (3) the version control, and (4) the starting of the tools. The system described in [203] uses EMACS as the top level interface. EMACS is used for editing the HTML pages. It spawns a browser to display the pages and it spawns a file management system to create the files and directories in the background. The file management system also calls the version management on the created files. While this system requires much more tool support, it also offers many advantages to the designer. First of all, the consistency between the HTML files and the file system is automatically achieved by the tools. For example, if the designer specifies a new file, he/she types the name into the EMACS. The name is used for creating the file via the file management system and an appropriate link is inserted in the HTML page. The second advantage is the automatic version control which is performed by the file management system when creating new files and by the EMACS when editing existing files. The system also provides a copying facility for templates. This supports the reuse of the templates which again makes the design data more consistent. The same organization is used for all setups (this is important when large databases should be modified automatically) and it saves time because a manual duplication is not required. The time is spend only once when creating a new template file.

The different levels make the HTML framework scalable. The designer can select the most suitable level of sophistication. The system can be upgraded if the design gets more advanced or if new versions of the system are designed, which are more complex. The underlying structure is a normal UNIX system. It uses the standards of the WWW. This means, all the software which is developed for that domain can be used directly in conjunction with the HTML framework. Expert users can use particular programs directly, i.e., the system is customizable. This is important due to the multi-disciplinary character of system design. In this context, a designer might be an expert in one domain and a novice user in another domain. The system must account for this by providing different levels of usage to fit for all users. These levels must interoperate with each other. Tools which are used directly in the UNIX domain by an expert user should not lead to inconsistencies when data are accessed via the HTML framework. This is often not the case with systems that perform a checking via a central database system. In these systems the tools must be invoked in a specific way. This forces expert designers to use specific design styles that may not suit them, making the whole design process less efficient.

The next section addresses the problems that have to be solved in this context. The following points are described:

- The organization of the HTML directories.

- The server and browser setup for enabling the start of design tools from the HTML pages.

- The automatic creation of the HTML environment via a generator. This keeps the design data and the HTML pages consistent. It automates the design processes.

- The integration of the complete system into the EMACS system. This provides a powerful front-end which links the HTML files that are created by the designer with the UNIX file structure.

- The version control and the groupware support. This part specifically addresses the unique challenges posed by the design applications. In this context it is very important to share the result data among all designers, since much CPU time was investigated to create the files. It differs from the typical groupware context of software development. For example, the compile times for software systems (except for very large ones) are often in a range that allows recompilation for every designer. Therefore, each designer can get his/her own private copy of the source code. The code is recompiled to obtain a working system. This contrasts with the quantitative analysis in system-level design. In this case, the exploration, for example, might easily require a total CPU time of several months. Hence, it cannot be redone for every designer.

The next section gives an idea of the general methodology. The description of the tool support for automating it is presented in [198].

ORGANIZING BY HYPERTEXT

In the HTML-framework context, the hypertext pages can be considered as a unique combination of a graphical user interface and a design documentation.

The links in the HTML pages are like buttons in a conventional user interface. The browser performs a certain action when a link is activated. Either it loads and displays a new page or it starts some helper application. The text and pictures in an HTML page are similar to pictures in a documentation. However, they are directly connected to the design data. Therefore, the updating of these pages is done during the normal design process. This avoids inconsistencies between the design documentation and the design data. Inconsistencies are often encountered when separate design documentations are used. It can happen that a designer performs certain actions on the design data without modifying the result tables of the documentation accordingly. In case of the HTML framework, the tools are started from the HTML documentation and the design data are referenced by the pages. Therefore the most recent data are automatically used in the pages.

In comparison to a conventional graphical user interface, the designer has a nearly unlimited choice of customizing the user interface without actually programming any tools. This feature is particularly useful for the top level configuration of the system. This part is highly application specific, which makes the development of an interface for all applications difficult. On the other hand, a direct starting of many different tools from the command line is only acceptable for an expert user. This would be difficult in a design context where many different designers with different background knowledge participate on a single project. Furthermore, the designer is not forced to use the graphical interface of the HTML pages. All design actions and all operations are still done by normal UNIX tools. Therefore the power of these tools [153] can be used directly. This might be necessary when more complex tasks have to be performed, for example, a restructuring of the design data base. This can involve operations which were not anticipated in the design of the HTML pages. The pages will normally provide just the basic commands that are useful for everyday design tasks. However, all operations fall back on the normal UNIX tool invocation. The designer has the choice to use the HTML interface or the UNIX shell. The appropriate choice depends on the task and on the experience of the designer. Both ways are completely compatible and can even be mixed. In addition to all these features, the HTML approach also provides the network support for the design. There is no further modification of the design documentation required. All the most recent data can be accessed via the WWW. For commercial applications the access might be restricted to particular design groups within a project. This section will outline the basic operation of the framework and the underlying WWW mechanism.

Fig. 6.1 shows the principal operation of the framework. The design exploration is used as an application example in this context. The links provide three major types of operations. They allow to setup the design step, they allow to update the results in case of modifications, and the results can be accessed without modifying them. The designer typically starts by creating a new directory for the design step. Next, the designer creates an empty HTML page for the organization. This page is called *README.html* in the context of the video processing design example. The designer enters the relevant context information for the current design step into the page. It is important for informing other designers about the applicability of the results. This avoids repeating similar design steps a couple of times due to information mismanagement. Next, the designer inserts a link to the tool for creating the setup specification of the exploration, e.g., *Architex*. The designer includes pointers to the expected results into the HTML page. This means, the HTML page is now an empty frame for the results that will be produced by the design step. The link on the setup tool is used to start the specification of the current design step. After finishing the editing of the setup, the design process can be started. This is done by calling a special link on the HTML page. The link is called *update*. It invokes the default command of the Makefile. This will start and control the execution of the design tools, in particular the time consuming measurement is performed in this way. No further actions of the designer are required in the measurement. It is done completely automatically. Otherwise a large measurement that runs through a week, or more, and needs manual attendance would be not manageable. This is why the shell interface of the analysis tools are so important. They allow to automate long running design tasks.

The tools generate two versions of the result files: the abstract data which are left for further usage by other tools and the rendered data which are referenced by the HTML pages. Typically, an inline image in GIF format is used to provide a quick overview of the data when scanning the result page. In addition, postscript versions of the data are created. They can be included into other documents and they provide a high quality displaying for investigating important data in more detail. This finishes the initial design step.

Other designers can now use all the results without invoking any design tools. They can use the update link to check if the data on the page are in a consistent state with the current design (this would be difficult in a conventional documentation). The page also contains the context information. It gives the background for evaluating the design data. It is also possible to include pointers to more background material into the page. This is often necessary in a complex design since not all members of the design team will have the same knowledge level for all different design steps. The references to the background material allow non-experts to become familiar with the material without spoiling the design data with lengthy introductions. This would have been necessary in non-hypertext documentations.

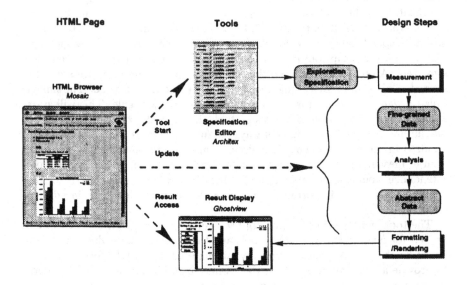

Figure 6.1.   Principal HTML framework scenario. The design exploration is used to ex-emplify the idea.

Normal designs are typically not done in a single pass. The designer will perform an analysis to investigate specific properties, e.g., different processor structures for a certain program. Based on the results of the investigation, it might be necessary to modify the program, for example, to increase the degree of parallelism. After performing the modifications, it is necessary to redo the analysis. This is not a problem in the HTML framework. The designer selects the update link. It checks the modifications and re-analyzes those parts that where effected by the modifications. Since the HTML pages refer to all design data by reference, they will immediately contain the new data after the updating of the directory is completed. The complete process works without any modification of the existing pages. Another modification might be the use of a different setup. For example, the designer chooses an initial design space for the exploration which turns out to be not suitable for the investigation. In this case the designer will click on the link of the setup tool to change the setup of the exploration. The setup tool will be called automatically with the correct specification file, this was specified when the link was inserted into the page. This means, if another designer wants to modify the setup he/she will directly follow the naming conventions used in the particular design context.

This shows how the HTML framework can be connected to tools of a particular design step. An important prerequisite for using the HTML browser as a customizable user interface is the invocation of tools via the browser. This requires a more detailed look at the basic mechanisms of the data exchange in the World-Wide Web (WWW) [126]. Fig. 6.2 depicts the overall structure of the client server communication in the WWW.

The client is an HTML-browser like *Mosaic* [189] or *Netscape*. This is the main interface for a user of the WWW. The client displays the HTML pages. In particular it performs the rendering of the pages. That is, it translates the formatting commands of the HTML into the appropriate fonts and indentations to be shown in the final page layout. But the WWW cannot only be used for exchanging HTML pages. It can be used for other page types as well. This allows to use links which refer to multimedia documents, like MPEG videos or audio data. Often the browser cannot provide all different display functionality that would be required for the various types of documents. Therefore it can start helper applications which perform the displaying of a document. An example is a PostScript text. In this case the browser can spawn a PostScript viewer like *ghostview* or *pageview*. The page is passed to the viewer which performs the appropriate *"displaying"*. This also applies to pages containing video data or audio data. In this case the browser would call the video or audio tool.

In addition to the displaying facility, the browser also provides the navigation on the links of the pages. This allows to jump back and fro among the links or

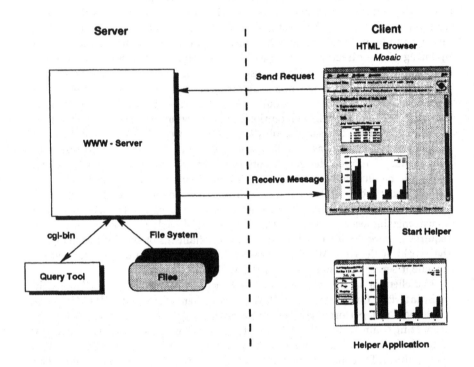

Figure 6.2.   The client/server structure of the World-Wide Web. The client can start helper applications to display multimedia messages.

to keep histories of visited pages. It makes accessing of hypertext documents easier for the user, because it allows to jump from some background material back to the main document. References can be followed to a certain degree without loosing the connection to the main document. The browser gets the data referred in a link either from the local file system (in this case it is used like a powerful file manager) or from a server in the computer network.

The server is an interface between a WWW database and a computer network like the Internet. The server is necessary to ensure the security of the database. For example, it can enforce authentication of a client before sending any data. The server also maps the file structure of the database to a file structure seen by the outside world. This is achieved by mapping the physical directories of the database into names seen from the outside world. The mapping allows to export just parts of a database or to use different names for the directories.

But the server cannot only return static files. It can also provide facilities for database queries. In this case it is normally not possible to prepare result files in advance due to the large number of possible results. A program must create the messages on the fly. Often the programs for answering the query are stored in a directory called *cgi-bin*. The server passes the query to a tool which performs, for example, a database search. The tool executes the search and formats the result as an HTML page. The server returns this page to the client. The approach permits to use tools for creating messages on the server side. The tool can use the query string as an argument.

The communication between the server and the client is done by a handshake protocol. If a client addresses a WWW server, the request is answered by a program called *http* daemon. Essentially, the daemon waits for requests from the outside world. For each request, it spawns a copy of itself to answer the request. The answer consists of a header defining the *application-type* of the body and the message body itself. The application-type forms the basis for describing the multimedia message bodies discussed above. That is, if different types of message bodies are allowed, the header is necessary to define the language interpreter that must be used for reading the message. Taking a PostScript message as an example, it would be next to useless, if this message is displayed directly as listing of the PostScript source code. An interpreter is necessary which translates the commands of the PostScript message into the page layout. In other words, a PostScript Viewer must render the message into a useful display-format for the user.

But the specific interpretation depends on the client. The server only defines that the message body is, e.g., in PostScript format. The type of PostScript interpreter to be used is defined by the client. This gives the client control over the tools that should handle the displaying. Clients can operate on the same application type in different ways. For example, a browser might directly inline a PostScript text or it can spawn an external tool. Clients can even chose not to display certain types of messages but rather to save the message body as a file where the user can invoke special tools by hand. This is the typical default behavior which a browser uses if an unknown application-type is encountered. The applications-types are defined in the MIME standard [23]. A *mailcap* file associates a helper application with each of the application-types. The mailcap file defines which external tool should be called when a certain application-type is encountered. If a browser reads a file from the local file system rather then getting it as a message from a server, the header information is missing. In this case it would be impossible to determine the appropriate application-type of the file. This missing information is filled by looking at the file name. Typically specific file extensions are used for the files. For example, a PostScript file normally uses the extension *".ps"*. A second mapping associates file extensions with the application types of the headers (the file is often called *extensionmap* or *mime.type*). This means when the browser fetches a local file, it will first determine the application type based on the file extension. Next, it will use the mailcap file to determine the helper application that should be used for displaying the file contents. Finally, the browser will spawn the viewer for displaying the contents of the file. Messages returned by a server are handled in the same way, except that the application type is determined from the header of the message. The body of the message is stored as a temporary file before a helper application can be called for displaying it.

A special application type are programming languages like *Csh*, *Sh*, or *Tcl*. In this case the body of the mail contains a program written in that language. A helper application would be an interpreter executing the program. This allows to send programs from a server to the client which will execute them. The user of a browser starts the execution by clicking on a link, i.e., the link in an HTML page works like a button for starting a program. In general this might by dangerous because a user should check a program before he/she executes it. This is not possible since the program will be directly executed when clicking on the application link. But in conjunction with a trusted server or the local file system this is not a problem. The user knows that no dangerous files will be returned by the server. In this case the user can enable the tool execution by associating the application-type with the appropriate interpreter of the programming language [59] (Normally the user will have two different browsers. One browser with program execution enabled. This browser is used in save environments and a second browser where program execution is disabled. It will be used for normal information retrieval from the Internet).

Sending the complete program would be too time-consuming for all but the most trivial programs. Therefore it is more efficient to send just the start code for the program. For example, a link for a design tool invocation will point on a *csh* script containing just one line: the command for starting the design tool. This means, the tool is started just like a normal UNIX program, however, the designer needs not type the command for starting the tool, rather, he/she uses the link in the HTML page. While this approach solves the problem of the tool invocation from the HTML pages, it leaves a second problem to solve. If the start up script is returned from a server as a static file, it cannot contain any arguments. This means all the design tools would be invoked without arguments. The designer has to load the appropriate files him-/herself into the tool, e.g., an analysis tool. This would be not very convenient because one major idea of the HTML framework was to support an interfacing to the design environment where knowledge about the underlying file structure is unnecessary. This means, the start script should contain the commands for starting the design tool with the appropriate arguments.

A simple approach, for solving this problem is to store all possible start up scripts directly in the design directories. However, this would spoil the directories with small scripts. It would be tedious to setup the script, and it would be difficult to change them if tool calls must be handled in different ways. A better approach is to use a server for creating the start up scripts on the fly. This can be achieved by using the database query facility described above. The tool arguments are passed as a query string to the server. The query tool takes these arguments to determine the appropriate start up script. It prepends the arguments found in the query string to the start up script. The complete script is returned to the client which executes the script to start the tool with the appropriate arguments.

Tools are invoked only at the client side. The server will not execute any design tools (i.e., a slow network is not a problem) . The server is only used for creating context specific start up scripts.

All the prerequisites for the HTML framework scenario of Fig. 6.1 are now in place. It is worthwhile to look at the update step in more detail because it differs slightly from a normal tool invocation. As outlined in chapter 5, the design steps will typically consist of a number of different tool invocations, i.e., a sequence of tool invocations must be performed rather than starting a single tool. Furthermore, it might be very time-consuming if, for example, certain parts of the measurement are repeated. This should only be done if the input to this part has changed. It would be unnecessary to repeat all the instruction-set simulation if only one additional data sequence should be included into the design data without modifying the existing parts of the analysis. Therefore the updating cannot be written as a normal program, because a program would repeat all updating steps all the time. A more refined approach is to use a *Makefile*. *make* [181] contains a list of possible targets that should be made.

Each target can depend on other targets or files. If the dependencies have changed, the target has to be updated. The Makefile contains the commands that must be executed to perform the updating of the target. That is, the Makefile contains not only the program code for the updating, but it also splits the updating into individual steps. The dependencies of each step are checked before executing the updating commands. This is an ideal concept for the design flow because each design step can be specified as a sequence of design sub steps which are only executed if necessary. This ensures that only the minimum amount of CPU time is spend on the updating.

make is typically called with the target to be updated as argument. A large number of targets can be specified. Each target might be considered as a specific command, where the processing steps are automatically adapted to the status of the current design. The Makefile will execute the design steps if it is necessary for making the design data consistent. The targets of the Makefile can be used as a kind of generalized call back facility for the HTML pages. As an example, the target *update* will perform whatever design operations are necessary to update the data in the current directory. This link can be included into every *README.html* page. More general, every command which requires more than the invocation of a single design tool can be implemented by a Makefile. The designer can write the complete design flow in the syntax of the Makefile. In addition, the GNU *make* program provides powerful pattern matching rules. For example, the conversion of PostScript files into GIF files can always be done in the same way. This can be stated as a pattern rule in a Makefile rule database. The *make* program will search this database and apply the rule whenever the command is not directly given in the Makefile of the current directory. In this way re-appearing rules can be stated in a compact way for the complete design environment. The *make* program will "know" how to convert PostScript files into GIF files. There is no need to restate such a step in every Makefile.

7 RESULTS

The thesis described a general codesign methodology for specialized processors. The different design tools were described and the configuration of these tools into complex design flows was explained. This section will present the results for a specific application spectrum: video compression. It is defined by the programs of Table 2.3 in section 2.1. The complexity of these applications reflects the characteristic challenges of designing high-performance video processing systems. The results are typical for the requirements imposed by video processing: a demand of high processing power and large memory bandwidth. But at the same time, the results also reflect special decisions and implementation constraints for the particular embedded system design. This means, they cannot be transferred completely to another design of an embedded video compression system. All these designs differ from each other. Exploiting these different constraints is one of the strong points of design automation for embedded systems. This makes it possible to tailor even a complex system to very specific constraints. It improves the overall performance / cost ratio of the system in comparison to more general solutions. Therefore the tools for measuring the results and the measurement strategy are more important than the specific values of the analysis results.

The presentation of the results follows the structure of the design flow. The first part is the quantitative analysis of the three different video compression programs. In particular, it shows how to use the sequence of analysis steps to define the structure of the processor. The second part describes the trade-offs which are possible in the cosynthesis phase. The focus is put on the IDCT (Inverse Discrete Cosine Transform) which is one of the most important functions in the video processing programs. Different algorithmic selections and hardware structures are compared to emphasize the trade-offs between the design domains of the embedded system (see Fig. 2.3 in chapter 1).

QUANTITATIVE ANALYSIS OF JPEG, H.261, AND MPEG

The three different video compression programs of Table 2.2 are used for the analysis. Each program is investigated with approximately 10 different data sequences. Standard sequences from research on video coding are used since they test the different aspects of the coding algorithms. The complete set of the results is available via World-Wide Web (WWW) [59]. The analysis of the three programs contains currently approximately 30000 results files. They are organized by about 900 HTML pages. Approximately 110 of these HTML pages were created by hand and 260 setup files were manually adapted from template files. These setup files were sufficient to create the remaining parts of the environment via the environment generator. In summary, the main part of the analysis is generated and organized automatically.

This section will discuss specific excerpts of the data to explain the analysis. The results are normalized in terms of a macroblock consisting of 16×16 pixel for the luminance component and 8×8 pixel for each of the two chrominance components. This allows to scale the data for different image formats according to the number of macroblocks given in Table 2.1. The normalization is not strictly necessary in case of the JPEG compression. But it facilitates the comparison of the results from the different compression programs. The first step is the requirement analysis.

Requirement Analysis

An outline of the steps for the requirement analysis is given in section 4.2. There are two main axes for the analysis: the refinement along the functions and the refinement along the different types of operations. The analysis will typically start with a refinement along the functions. Table 7.1 compares the operation count for the complete program and the most demanding functions. The data for the different compression programs and different data sequences are given. The total operation count depends slightly on the different data sequences. This will be investigated in more detail below. But the implementation of the different programs makes an even more significant difference. For example, the *mpeg_play* program implements the decoding of the MPEG-1 algorithm. Decoding of this algorithm is more complex than the decoding of the baseline JPEG. Despite this fact, the *mpeg_play* consumes only one fifth of the operations required by the *PVRG-jpeg*. In both cases the programs where compiled with all optimizations of the GNU/MOVE compiler. This means the factor five improvements obtained by the *mpeg_play* program were achieved by a sophisticated optimization of the software implementation. For example, the *mpeg_play* avoids the storing of intermediate results in arrays. It was described in section 4 that such an optimization can lead to a significant reduction of the memory traffic. Section 7.17 will investigate more optimizations by looking at the IDCT as an example.

The improvements show how important the analysis and the accompanying knowledge based optimizations are. For comparison, in [117] an improvement in the order of 20% was achieved by an optimal scheduling of basic blocks. Exploiting the knowledge of the designer gives nearly two orders of magnitude higher improvements. Even more important, both optimizations can be combined to take advantage of the designers knowledge and the time-consuming automatic optimizations. In this case, it is important to restrict the automatic optimizations to the most important functions. This keeps the CPU times for the optimization in realistic ranges. For example, the *mpeg_play* program has a total of 150 functions but 60% of the operations are consumed by the 3 most important functions. This means that the locality of the program execution can be exploited for the optimization. Of course, the scheduling of this part is still NP-complete, but the size of the problem is significantly reduced. In this way even optimal algorithms can be applied to complex designs [118]. The quantitative analysis allows to restrict them to a small part of the problem where the gain from the optimal algorithms is highest.

The optimization of the different programs is important for improving the system efficiency. But the operation count data are random variables which depend on the video data sequences. As discussed in section 2, the designer must ensure that a typical ensemble of data sequences is used. Since it is impossible to investigate all possible image sequences, it is necessary to start the requirement analysis with a large number of sequences. The behavior of some of these sequences is summarized in Table 7.2. The selection of the first ensemble is based on the data sequences that are typically used for analyzing video compression algorithms. These sequences test the various image contents: smooth surfaces, different textures, different types of motion, etc. The contents in turn influence the processing within the coder. In particular, the selection of the macroblock type depends on the contents. Three main macroblock types are distinguished in the MPEG-1 coder: (1) Intra-Macroblocks (I-MB) where

Table 7.1.   Most demanding functions in the video compression programs of Table 2.3.

Demanding functions

Decoding Program	Input Sequence	Operations / Macroblock	Important Functions		
			Rank	Name	% of Ops
PVRG	lena.jpg	47754	1.	ChenIDct()	25.92%
			2.	DecodeHuffman()	13.36%
			3.	megetb()	11.18%
-JPEG	barb.jpg	49568	1.	ChenIDct()	25.06%
			2.	DecodeHuffman()	14.25%
			3.	megetb()	12.03%
PVRG	sflower.p64	52525	1.	LoadFilterMatrix()	26.64%
			2.	ChenIDct()	17.63%
			3.	mgetb()	14.34%
-p64	stennis.p64	29091	1.	LoadFilterMatrix()	22.58%
			2.	ChenIDct()	16.76%
			3.	WriteBlock()	11.88%
mpeg	flower.mpg	9402	1.	j_rev_dctOpt()	22.18%
			2.	ReconPMBlock()	19.19%
			3.	ReconBiMBlock()	16.73%
_play	tennis.mpg	10198	1.	j_rev_dctOpt()	26.29%
			2.	ParseReconBlock()	23.70%
			3.	ReconBiMBlock()	20.77%

only the current frame is used for the coding of the block, (2) Predictive-Macroblocks (P-MB) where only a one-sided prediction is used for the coding the macroblock, and (3) Bidirectional-Macroblocks (B-MB) where backward and forward prediction is used. The distribution of these macroblock types for the image sequences of Table 7.2 is depicted in Fig. 7.1. This is an example which shows the importance of a careful system analysis, as discussed in the following.

Bidirectional prediction provides the best prediction quality because the current block is interpolated from an image block in a future frame *and* an image block in the past. This means the changes in both of these frames are considered in the prediction. As a disadvantage, both frames are necessary for the reconstruction of the block in the decoder. That is, two inverse predictions must be performed for reconstructing a B-macroblock. A naive approach to a video processor design would thus assume that the B-macroblocks are the worst case and thus should be avoided to keep the system requirements low. The results of Table 7.2 show, however, that the contrary is true. As depicted in Fig. 7.2, those sequences which use a large amount of B-macroblocks, (*Reds-Nightmare.mpg* and *anim.mpg*, use the lowest overall operation count, up to 20% less than the other sequences.

Table 7.2. Most demanding functions in the *mpeg_play* program. Intra coded macroblocks(MBs) are denoted by I-MB, predictive coded (either forward or backward) MBs by P-MB, and bidirectional coded MBs by B-MB.

mpeg_play: **Demanding functions**

MPEG File	Number of Macroblocks(MBs)			Most Demanding Functions	% of Operations	Operations per MB
	I-MB	P-MB	B-MB			
Reds-Nightmare.mpg	28859	141740	183325	ReconBiMBlock()	33.90%	7035
				j_rev_dct()	17.06%	
				ParseReconBlock()	12.48%	
anim.mpg	1437	1421	3508	ReconBiMBlock()	31.76%	7618
				ParseReconBlock()	20.81%	
				j_rev_dct()	18.80%	
c4.mpg	1992	8798	7143	j_rev_dct()	23.97%	10929
				ParseReconBlock()	19.18%	
				ReconBiMBlock()	17.93%	
flowcr.mpg	4847	25077	14638	j_rev_dct()	22.18%	9402
				ReconPMBlock()	19.19%	
				ReconBiMBlock()	16.73%	
tennis.mpg	9238	17019	21594	j_rev_dct()	26.29%	10198
				ParseReconBlock()	23.70%	
				ReconBiMBlock()	20.77%	

The low overall operation count results from savings that are made in other functions of the decoder. These savings are made possible by the good prediction. It makes the removal of redundancy very efficient so that many of the input coefficients to the inverse transformation are zero. This in turn allows to skip operations in the IDCT. Hence, the operation count for the IDCT function ($j\_rev\_dct()$) is reduced. In the same way, the operation count for the variable length decoding and the inverse quantization is lowered. In summary, the savings made by these other operations more than compensate the additional operations required for the reconstruction of the B-macroblocks. This means a naive guessing of the complexity of the coder is not enough. Even worse, if a designer uses the basic knowledge about the individual functions to estimate the complexity of the complete system, this can even lead to completely unexpected results. For instance, if the designer would use the sequences with many B-macroblocks as "worst case" test scenario, this would lead to a completely wrong design which under-estimates the required performance by as much as 20%. It shows that the inter-dependencies between the different parts of a complex system are often not obvious. Therefore the measurements are necessary to investigate the behavior of the system.

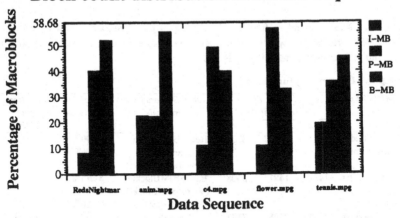

Figure 7.1. Distribution of macroblock types in different MPEG data sequences. *RedsNightmare.mpg* and *anim.mpg* use the highest number of B macroblocks.

Another example, where it is important to consider different operating conditions of the decoder, is shown in Table 7.3. The operation count for different compression ratios is depicted. The sequence *lena25.jpg* uses the highest

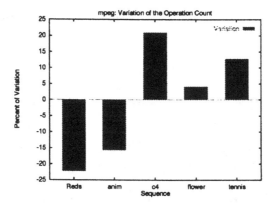

Figure 7.2. Variation of the total operation count for different MPEG sequences when decoded by the *mpeg_play* program.

compression ratio, that is, the amount of data that needs to be stored or trans-mitted is the lowest (see section 2). A high compression ratio also reduces the operation count. The reduction of the operation count in comparison to the sequence with the lowest compression ratio is shown in the row *reduction*. Up to 28% of the operations are saved when higher compression ratios are used. In particular, the amount of operations for parsing the coded data stream is reduced. This changes the ranking of the operations. While the variable length decoding (*DecodeHuffman*) is not very important when high compression ratios are used (*lena25.jpg*), it becomes more important when the compression ratio is decreased (*lena.jpg*). With low compression ratio it is the second most im-portant function of the JPEG decoder. It consumes 13.36% of the operations. The interpretation of these data depends on the application context of the em-bedded video compression system. If the system is designed for supporting all kinds of operation conditions, even high quality images with low compression, the designer must take into account that the requirements increase if a lower compression ratio is used. On the other hand, if the system is used only in conjunction with low bitrate channels, for example, in a hand-held system, this means, the designer can take advantage of the lower operation count. In the latter case, the variable length decoding is less important in comparison with the storing of the reconstructed image. Again, the data show that a careful analysis is important for designing the system, without using oversized compo-nents or introducing unexpected bottlenecks, such as a slow Huffman decoding of the compressed data. Furthermore, the interpretation depends on the application context. This means, the same data might lead to completely different design decisions. Therefore the designer is indispensable for interpreting the analysis data.

Table 7.3. Most important functions of the JPEG decoder when different compression ratios are used. *lena25.jpg* uses the highest compression and *lena.jpg* the lowest.

JPEG: Demanding functions

Program Data	Input Data Sequences			
	lena25.jpg	*lena50.jpg*	*lena75.jpg*	*lena.jpg*
Operations / Macroblock	34520.6	36316.9	39073.7	47709.4
Reduction	-28%	-24%	-18%	0%
1. Function	ChenIDct	ChenIDct	ChenIDct	ChenIDct
% of Operations	35.82	34.05	31.65	25.92
2. Function	WriteXBuffer	WriteXBuffer	WriteXBuffer	DecodeHuffman
% of Operations	14.08	13.38	12.44	13.36
3. Function	IZigzagMatrix	IZigzagMatrix	DecodeHuffman	megetb
% of Operations	8.92	8.48	7.99	11.18

Up to this point the discussion has focused on the total operation count and on the distribution of these operations among the different functions. The second analysis axis is the distribution of operations among the different operation types. Section 4.2 has investigated the conditions for mapping several different application programs onto a processor. A necessary condition in this case is the similarity in the instruction mixes. The maximum difference of the instruction mixes determines the throughput reduction that will be achieved even under the best scheduling conditions. Fig. 7.3 compares the instruction mix for the different programs and the data sequences of Table 7.2. The requirements are very similar, especially for the demanding operation types like load, store, and mult. This means the data path requirements do not differ in principal. The programs are good candidates to be processed by a single processor. This gives an advantage for a specialized processor in comparison with a general-purpose processor. The latter must be designed to support all kinds of different applications while the specialized processor is tuned for a particular application domain. However, a restricted application domain does not necessarily ensure that the programs impose the same requirements on the underlying hardware. The measurement confirms this. In case the condition is not fulfilled, the designer can change the most important functions to achieve a more uniform instruction mix.

On the other hand, the instruction mix analysis of the different program shows the independence of the quantitative analysis from the specific semantics of the programs. The analysis is an abstract way of representing the system. This is unlike more detailed models, for example, a CDFG (Control Data Flow Graph). These models reflect the details of the implementation, however, they are also completely changed when the syntax of the program is changed. The quantitative analysis extracts a more global view that depends less on the details of the implementation and more on the general requirements of the application. This is confirmed for the programs of Table 7.2 which even impose different total requirements, i.e., they differ significantly in their total efficiency, but the impose similar requirements on the data path structure.

Instruction Mix of Different Algorithms

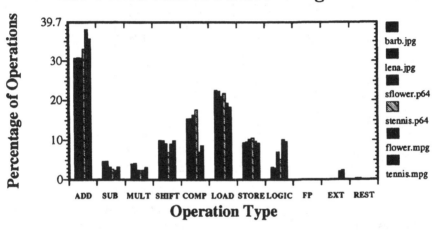

Figure 7.3.   Instruction mix of the different video compression programs in Table 7.1.

As shown above, the compression ratio of the programs also influences the operation count. Therefore it is important to compare the dependencies of the instruction mix when using different compression ratios. The data for the JPEG decoder of Table 7.3 are summarized in Table 7.4. Again, the instruction mix is approximately constant. Only a slight influence of the instruction mix is visible since the variable length decoding functions get more important. This storing of the reconstructed image and the IDCT are less important when using low compression ratios. As a consequence, the percentage of multiplications and store operations is slightly reduced. The changes are not very significant. This means, the JPEG decoding at different compression ratios can be done on the same type of processor, but the time per macroblock will increase when a lower compression ratio is used. While these small changes require no significant changes in the structure of the data path, they show that the instruction mix data reflect the changes in the behavior of the programs. This means the instruction mixes are well suited to analyze the dynamic behavior, if different data sequences or different working conditions are used.

Table 7.4. Instruction mix of the JPEG decoder when data sequences with different compression ratios are decoded.

jpeg: Instruction mix

Instruction Type	Input Data Sequence			
	lena25.jpg	lena50.jpg	lena75.jpg	lena.jpg
ADD	32.23	31.98	31.63	30.81
SUB	4.77	4.75	4.72	4.64
MULT	5.74	5.46	5.07	4.16
REST	0.47	0.44	0.41	0.34
EXT	0.00	0.00	0.00	0.00
COMP	15.85	15.77	15.66	15.41
SHIFT	9.10	9.24	9.42	9.87
LOGIC	0.84	1.21	1.72	2.94
FP	0.00	0.00	0.00	0.00
LOAD	20.10	20.49	21.02	22.29
STORE	10.90	10.66	10.33	9.54

The requirement analysis has focused on the complexity of the different video compression programs and on the changes,that result from different input sequences. Little assumptions about the hardware were used in this part of the analysis. This allows to focus the discussion on the algorithms and on the software implementation. The instruction-set of the hardware was fixed and a load/store processor architecture was used. The next section will address the influence of the processor hardware structure in more detail. For this purpose it is useful to extend the requirement analysis to the requirements imposed on the control unit. This is shown in the branch statistics of Table 7.5.

The branch statistics allow to estimate the influence of the control flow dependencies on the performance of the processor. A modern processor relies on a pipelined processing of the instructions (see section 2.4). New instructions are fetched while the processing of the current instruction is still in progress. This scheme gives good performance, if no control flow changes are encountered. Direct conditional branches can have two possible paths for the control flow. The chosen path depends on the result of the branch condition. This is often determined at a later stage of the pipeline. New instructions are fetched without knowing the correct path. The path that is selected for fetching the instructions

Table 7.5.   The branch statistics of the different video compression programs when decoding the data sequences of Table 7.1.

Branch Statistics for Different Video Decoders

Branch Data	Input Data Sequence					
	barb.jpg	lena.jpg	sflowg.p64	stennis.p64	flower.mpg	tennis.mpg
Ops/block	4.9	5.0	4.2	4.0	10.1	8.1
Successors/block	1.7	1.7	1.6	1.7	1.7	1.7
No of branches	7603.9	7353.6	8546.9	5105.6	655.2	879.4
Forward taken	34.94	35.62	37.52	34.22	45.57	43.58
Forward not taken	21.11	19.42	15.99	9.64	42.62	47.32
Forward indirect	0.00	0.00	0.00	0.00	0.01	0.00
Backward taken	40.41	41.51	43.22	52.27	9.78	7.58
Backward not taken	3.53	3.44	3.27	3.86	2.02	1.52
Backward indirect	0.00	0.00	0.00	0.00	0.00	0.00

is either selected by default (e.g., branches are assumed to be not taken), or a branch prediction is applied to select the path. In this case, the selection is based on the previous behavior of the program at that point. It assumes that the history of the branch conditions can be used as a prediction for the future.

A wrong prediction of the branch is typically handled by canceling the operations that were wrongly fetched. The pipeline is flushed, and the instructions along the correct path are fetched instead. This degrades the performance of the processor because all instructions that are canceled do not contribute to the processing of the algorithm. They can be regarded as idle cycles. The number of idle cycles that are produced in this way depends on the latency between the fetch of the branch instruction and the calculation of the branch condition. A high degree of pipelining typically leads to a large latency, i.e., a large number of idle cycles. This means a high degree of pipelining should not be used if the branch prediction is not good, or if the average number of instructions in a basic block is low.

Table 7.5 shows these data for the three different video compression programs. The programs differ slightly in their behavior. The *mpeg_play* program uses loop unrolling and high optimization, especially in the IDCT (see section 7.17). This has two consequences: (1) the average length of the basic blocks is increased and (2) the number of backward branches is reduced. Both consequences stem from the fact that a branch at the end of the loop often jumps back to the beginning of the loop. There are often large numbers of taken backward branches. These branches can be found in the other two programs. The remaining branches in the *mpeg_play* program are mainly forward branches which are taken or not taken with approximately the same probability. Some of the optimizations in the *mpeg_play* program use branches to save operations. This is used, for example, in the IDCT. There are often input data which contain zeros. Branches are used to check for these zeros. Operations are skipped, if zeros are encountered. This makes the branch predictions difficult because the branches are highly data dependent. The outcome of previous branches is not very representative for the behavior of the next branch. This means a high degree of pipelining should be avoided for the *mpeg_play* program.

The situation for the two other programs is different, however, a deep pipelining should be avoided as well. In this case the loops contribute more to the number of branches. Therefore approximately 40% to 50% of the branches are taken backward branches. This simplifies the prediction, since backward branches can be predicted as taken (This can be either used as default scheme or it will make a real branch prediction more efficient). However, the average length of a basic block was decreased. This increases the likelihood of pipeline hazards and thus counter-effects the better branch prediction possibilities. In summary, the degree of pipelining in the processor should be low for all three applications.

The analysis of the branch statistics is necessary for selecting the correct setup of the design exploration. As described in section 2.4, the functional units of the processor might be pipelined. The degree of pipelining influences the scheduling in the design exploration. This means, a realistic pipelining degree of the functional units must be selected for the exploration setup. Using a large degree of pipelining might lead to a large number of idle cycles because the scheduler must take into account the latency of the functional units before a result can be used. At the same time, the clock frequency of the hardware can be increased (see equation 4.2). Since the branch statistics already showed that a large degree of pipelining is not useful for the considered application spectrum, this knowledge can be used to restrict the time-consuming exploration to a useful range. For the exploration in the next section the following latencies are selected:

Multiplier: uses three pipeline stages

Memory: uses a latency of two

All other functional units: use a latency of one cycle.

This setting balances the pipelining for the different elements of the processor. A trade-off was made in favor of a lower degree of pipelining. This means a modest clock rate is achieved, but the overall utilization of the functional units is kept high, because the scheduling efficiency is high and the branch delays are small.

Design Exploration

The previous section has focused on the results of the requirement analysis. This analysis step requires only little knowledge about the underlying processor structure. The purposes of the requirement analysis are twofold. First of all, it allows to improve the efficiency of the software implementation and the algorithms. This is achieved by optimizing the most important functions of the application programs. The second purpose is the extraction of a useful setup for the design exploration. For example, the selection of the pipelining degree for the functional units taking into account the results of the branch statistics. The ensemble of data sequences, which is used for the exploration, is based on the investigation results from the requirement analysis (i.e., the requirement analysis and the knowledge of the algorithms allow to select a representative image ensemble). In summary, it is a crucial step for improving the implementation of the video compression system, but the requirement analysis is not sufficient for designing the processor structure.

In particular, the knowledge about the scheduling efficiency for a particular processor framework is not included. The requirement analysis data provide not much information about the available instruction level parallelism, or about the required size of the register file. This information is crucial for determining the actual processing time of the algorithms. For example, an algorithm with few operations might seem favorable in the requirement analysis. But if the reduction of the operation count is achieved by a large number of control and data flow dependencies, this can lead to a low degree of instruction level parallelism. This would make the algorithm less useful for complex processors with many functional units. Additional analysis data are necessary. The data must reveal the performance that is achieved for a particular hardware structure. These data are provided by the design exploration. The data are presented in this section. The VLIW (Very Long Instruction Word) architecture of section 2.4 is used as an architecture framework. The MOVE [80] scheduler is used for the parallelization of the opcode. The processing times are normalized in cycles per macroblock. It provides a seamless way to compare results from different semiconductor technologies and different image sequences. The section exemplifies the use of the exploration results for gradually fixing the processor structure. The approach is subdivided into three major steps (see section 4.3). Each of these steps investigates the performance, that is achieved when putting certain constraints on the processor resources. The first step is the disjoint exploration. It investigates the resource constraints for each unit type individually. The second step is the combined analysis where restrictions are imposed on several different unit types at the same time. This step analyzes the effect due to the combination of resource conflicts for several different units.

As an example, the disjoint exploration might show the usefulness of a large
number of adders. But the efficiency of these adders depends on the availability
of the input operands for the additions. The input operands are typically provided either by the register file or by the load units (in case of data coming from
the memory). This means the large number of adders depends also on the large
number of memory ports or on a large register file. The combined exploration
investigates the first case. The inter-dependence of the resource constraints
among the different functional unit types. The second case, the adaptation of
the register file is investigated during the register file exploration. This step
follows the combined exploration. It adapts the size of the register file and the
number of input and output ports. All these parameters depend on the number
of functional units and on the different algorithms. This is detailed below.

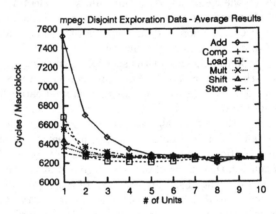

Figure 7.4.   The results of the disjoint exploration for the *mpeg_play* program.   The
average over 13 different image sequences is taken.

The starting point of the exploration is the disjoint exploration. An example
of the results for the *mpeg_play* program and an average of 13 different data
sequences is given in Fig. 7.4. Each curve depicts the performance when
restricting a particular unit type. The unit type is given as the parameter
of each curve. Only the indicated unit type is restricted while all the other
unit types of the processor framework are left unconstrained. The number
of these unrestricted unit types is set to a very large value. Therefore, the
scheduler always finds a free unit for these unit types. This means the cycle
times represent bounds on the performance for processors with a particular
number of functional units. For example, the point $(2units, 6700\frac{cycles}{Macroblocks})$
on the adder curve is a lower bound on the cycle time of *all* processors with 2

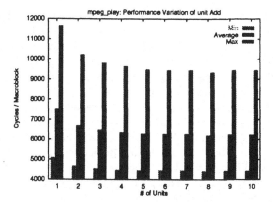

Figure 7.5. The performance variation when using different image sequences as data input. The results of the *mpeg_play* program and the adder unit are shown. The minimum, average, and maximum cycle count is shown for an ensemble of 13 image sequences.

adders, when the VLIW framework of section 2.4 and the MOVE scheduler are used. In this way each of the curves in Fig. 7.4 can be regarded as a hull on the performance. This is useful for restricting the design space because typical programs have only a limited degree of parallelism. Providing more functional units than available parallelism does not help because the dependencies of the operations prevent further parallelization. In case of the *mpeg_play* this is achieved at around $6250\frac{cycles}{Macroblock}$. Even using very large numbers of functional units, i.e., up to ten units, does not give any performance improvements. In case of adders, 5 adders give approximately the same performance bound as ten adders. Therefore the next explorations are restricted to ranges where the performance changes are more significant. In this way the exploration subspace for the combined exploration is restricted to an exploration of $\{1, \ldots, 5\}$ units for the adder and $\{1, \ldots, 3\}$ units for all other unit types.

In addition to specifying the setup for the combined exploration, the designer can use the results in a number of other ways. He/she can use them to modify the algorithmic properties [118]. This completes the investigations of the requirement analysis because the data dependencies between the operations, the latencies of the pipelined units, as well as the resource constraints on demanding unit types are taken into account. The data can be used to distinguish between algorithms which are more favorable for processors with a small degree of instruction-level parallelism and between algorithms that can be used on larger processors. This is investigated in more detail in section 7.17.

The disjoint exploration can also be used to determine the performance variations when processing different data sequences. This was investigated partly during the requirement analysis. A variation of the operation count in the order of 20% was observed (Fig. 7.2). But the operation count variations do not consider the dependencies among the operations. This might change the observed variations. The disjoint exploration can be used to investigate this behavior of the different sequences. A large number of sequences can be investigated since the number of points in the disjoint design space is comparatively small (see section 4.3). The results for 13 different data sequences and the *mpeg_play* program are shown in Fig. 7.5. The graph depicts the performance variation for processors with different numbers of adders. The x-axis indicates the number of adders in the processor. The y-axis shows the corresponding operation count. Three values are shown for each number of adders: the minimum cycle count. This indicates the processing time for a data sequence that required the minimum number of cycles. Next, the average cycle time is shown. This can be considered as the typical processing time that was achieved for the selected ensemble of image sequences. This is the adder curve shown in Fig. 7.4 that is used to compare the different unit types. The last value is the maximum processing time. It is the worst case performance for a data sequence from the ensemble. The variation of the processing times are slightly increased in comparison to the variation measured in the requirement analysis.

Finally, the disjoint analysis is used for a detailed investigation of the most important functions. They are typically identified and improved during the requirement analysis. In addition, they might be distributed on different processors. In this case it is often useful to combine those functions which allow a high degree of instruction-level parallelism, i.e., those functions which gain most from a larger processor. An example is the *j_rev_dct* function shown in Fig. 7.6. The principal appearance is similar to the overall performance improvements in the disjoint exploration of the complete *mpeg_play* program. However, the gains from increasing instruction level parallelism are higher. The performance nearly doubles if a larger processor is used for the *j_rev_dct* function.

A similar finding holds for the PVRG-*jpeg*. In this case the two most important functions are the *DecodeHuffman* function and the *ChenIDct* function. The disjoint exploration data for these functions are shown in Fig. 7.7 and Fig. 7.8, respectively. The two functions have different characteristics. The *DecodeHuffman* function reverses the variable length coding of the transmitted image data. This means, the bit stream must be parsed and the decoding of the next code word requires the translation of the current code word. The process is sequential, if implemented on a normal processor. In most cases, many control flow and data dependencies exist. The *ChenIDct* function is different.

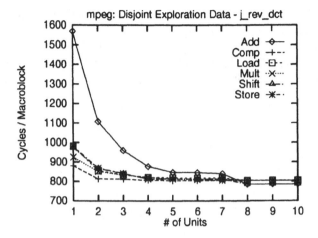

Figure 7.6.   The disjoint exploration results of the *j_rev_dct* function.

It contains mainly data processing and loops with fixed loop bounds. Therefore it contains more possibilities for parallelizing operations.

This is reflected by the disjoint exploration results of Fig. 7.7 and 7.8. In case of the *DecodeHuffman* function, only a small improvement is obtained when using more functional units. The improvements are mainly achieved by increasing the memory bandwidth, i.e., providing more load and store units. Furthermore, the possibilities for parallelization are immediately saturated. Except for the store units, there are no more improvements when using more than two units. This shows the sequential nature of the parsing process in the *DecodeHuffman* function. The situation is different for the *ChenIDct* function. In this case there is a larger overall improvement when increasing the available instruction-level parallelism. The performance is increased by approximately 40% when using more functional units. A larger degree of parallelism is possible. This varies slightly with the different unit types. The performance does not improve when using more comparators. But increasing the number of adders or multipliers gives a significant speed up. The interpretation of these speed up results must take into account the costs of the different unit types. For example, providing more adders or shifters will not increase the overall system costs very much. The gate count for these unit types is comparatively small. Increasing the number of multipliers will result in a very significant increase of the gate count for the processor data path. Typically, this is not useful, if the performance improvements are low. The most expensive improvement is the larger number of load/store units. This requires to increase the overall memory bandwidth and it will require multi-ported caches which are typically slower than single-ported caches. Hence, a careful trade-off is required to ensure that the additional costs of the system are justified by the achieved performance improvements. Once again, the designers knowledge is necessary for interpreting the results of the analysis.

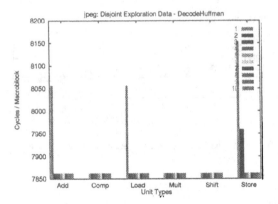

Figure 7.7.   Disjoint exploration results of the *jpeg* program. Only the exploration of the *DecodeHuffman* function is shown.

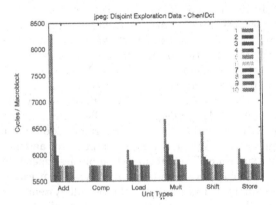

Figure 7.8.   Disjoint exploration results of the *jpeg* program. The cycle counts for the *ChenIDct* function are shown.

The different characteristics of the *DecodeHuffman* function and the *ChenIDct* function are obvious from the data of the disjoint exploration. These characteristics provide a first criterion for distributing the functions in a heterogeneous multiprocessor system. The designer has two major informations for selecting a suitable task distribution. First, the instruction mixes from the requirement analysis can be used to determine the similarity of functions or programs (see section 4.2). But these data do not take into account the data dependencies among the different operations. An ideal scheduling was assumed for the evaluation of the throughput reduction. This missing information is provided by the disjoint exploration. It can be considered as an extension of the instruction mix analysis. In this case the dependencies among the operations and the latency of the operations in pipelined functional units is taken into account. In the example above, the *DecodeHuffman* function will not fulfill the ideal scheduling condition, since too many data dependencies hinder the parallelization. On the other hand, the *ChenIDct* function or the *j_rev_dct* function are more amenable for the parallelization. They have similar improvement characteristics when increasing the number of functional units. Hence, they are good candidates to be mapped onto the same processor in a heterogeneous multi-processor system. In other words, the instruction mix data of the requirement analysis give performance bounds and the disjoint exploration shows how tight these bounds are.

The disjoint exploration is advantageous for these investigations due to the reduced complexity (see section 4.3). Only one processor is explored for each unit number. This reduces the overall size of the exploration so that many different data sequences or functions can be investigated. The missing information is the performance influence of the combined restriction of several different unit types. As mentioned above, the adder performance might depend on the number of load / store units. This makes the analysis more complex because all possible processor constellations have to be investigated. The disjoint exploration results are used to restrict this demanding exploration. A subset of the original image sequences is chosen and the unit ranges are selected so that they cover the interesting parts of the design space. The combined exploration will investigate this important subset of the original design space. Based on the results, the structure of the processor data path is selected.

The selection process is described for the results of the combined exploration of the *mpeg_play* program and the *flower.mpg* data sequence (Fig. 7.9). The picture shows the first two steps of the unit selection. The figure shows the global summary view, the corresponding unit views of the first level, and the sub-summary view after fixing the first unit type of the processor data path. These are the first steps of the analysis procedure described in section 4.4.

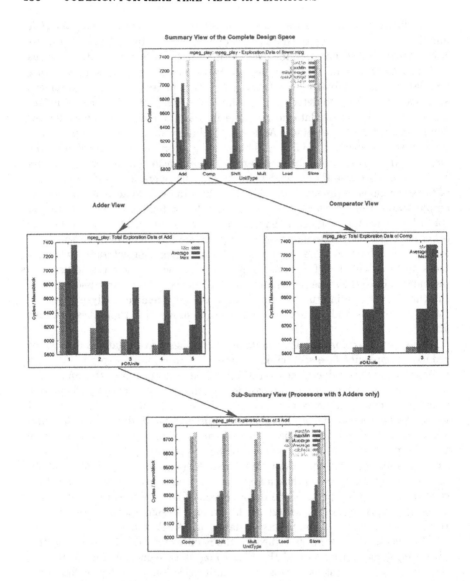

Figure 7.9. Selection of the processor structure in the combined exploration. The data for the *mpeg_play* program and the *flower.mpg* data sequence are shown.

The summary view shows the different unit types on the x-axis and the corresponding cycle times on the y-axis. The cycle times are normalized w.r.t. a macroblock. For each unit type six values are shown as bars. Pairs of bars represent the bounds (upper and lower bound) for the minimum, average, and maximum values of the corresponding unit view. The variation between the leftmost pair of bars for each unit type is the upper and lower bound of the minimum cycle time in the unit view. The variation of the bar pairs reflects the performance variation of the bounds when changing the number of units for the indicated unit type, i.e., the *intra-unit* variation. The *inter-unit* variation is the variation between corresponding bounds, for example, the upper bound on the minimum cycle time (the second bar from the left in each group) and the upper bound on the maximum cycle time (the rightmost bar in each group). The inter-unit variation is an indicator of how much the performance is influenced by the constellation of the other unit types.

Two different cases can be distinguished in Fig. 7.9. First of all, unit types with high inter-unit variation. Examples are the comparator, the shift units, etc. In this case the performance is more or less independent from the number of units which are provided for these operations. This is obvious from the comparator view presented in Fig. 7.9. It shows that the average performance of processors with one or three comparators is approximately the same. But there is a large difference between the minimum cycle time of processors with one comparator and the maximum cycle time of a processors with one comparator. This means, the performance depends much more on the other unit types. In summary, the specific number of units with low intra-unit variation is not very important for the overall performance. There are other unit types which are much more crucial. These are unit types like add and load which have a high intra-unit variation. In this case the performance of the processor depends very much on the number of functional units provided for each of these unit types. The most important unit type is the adder, the one with the largest intra–unit variation, as shown in the adder view of Fig. 7.9. For example, the best-case performance of processors with only 1 adder is in the same range as the worst-case performance for processors with two adders. Hence, it is favorable to use more than one adder in this case. However, the balancing between performance gain and cost increase requires a careful trade-off from the designer. Many different factors must be considered. In case of the adder, the gate-count costs for the adders are small. This means more adders can be used without leading to unacceptable sizes of the data path. This is different for demanding units like multipliers. But a large number of adders requires also a larger interconnection network. This means, more area is occupied by busses and it can lead to a slower processor design, since a large interconnection network will have a higher delay. The clock rate must be reduced. As a rule of thumb, the unit number immediately before the saturation starts is the most preferable one. In the example, 3 adders are required. More than three adders give little additional benefit, but will produce the disadvantages described above. The designer will therefore focus on the design space of processors with three adders, a subset of the original design space.

For this subset, the situation of selecting unit number is similar to the selection in the complete design space. Again a summary view is required to determine the most important unit type. This time the summary view contains only those processors with three adders. The adders are already determined in the data path. Thus the adder does not appear as selectable unit type on the x-axis of the sub-summary view. The dimension of the original design space is reduced by one. The intra-unit variation for this reduced design space is highest for the load units. The designer will select the suitable number of load units from the load unit view. It is calculate for the design space of processors with 3 adders only. In this way the design space is gradually restricted to the most useful processor structure for the particular embedded system design.

The main advantages of this approach, in comparison to a selection of the units by some optimization algorithm, is the controllability by the designer. The possible performance improvements or losses are directly visible to the designer. The designer can select unit types in almost arbitrary ways. The selection strategy can be changed within the selection process, if other criteria are more important than the performance of the system. For example, the designer needs not follow guidance of the intra-unit variation. Demanding units might be fixed without using the number of units that give the best performance. This might apply, for example, to the load units. If the bandwidth of the main memory is the major cost factor of the embedded system, the designer can start the selection process with the load units instead of considering the adders. This overriding of the standard selection procedure can be applied at any point of the design process. It gives the designer much more freedom and control than an optimization by an algorithm (such as integer linear programming). The major disadvantage of the optimization algorithms is the poor control of the selection process. Only the cost parameters can be used to change the selection. However, small changes of the cost parameters might lead to completely different results. It is often difficult to consider special constraints in the optimization process. Using the recursive selection of units, supported by the visualization of the analysis results, the optimization trajectory is directly controlled by the designer.

In summary, the combined exploration allows to determine the general structure of the processor data path. The analysis takes into account the properties of the algorithms, e.g., the available parallelism, the required operation types, the necessary number of memory ports, etc. The designer completes this performance information with the knowledge about the costs of the units and the specific design constraints. The visualization of the analysis results is the central element for achieving this kind of designer centered approach. In case of the video compression processor design, the adder was the most important resource for achieving high performance. However, a large number of adders requires high memory bandwidth. A typical situation for the video compression algorithms, since they operate on large streams of data. Due to the high system costs, which occur when providing high memory bandwidth, it is important to keep the number of load / store units at a minimum (see section 4.2). Balancing both contradicting requirements leads to a selection of 2 load units, 1 store unit, and 3 adder. The remaining functional units are less critical for the overall performance. An exception is the multiplier. In this case it would be advantageous to use more than one multiplier. However, the multiplier is very demanding in terms of the gate count. It will dominate the complexity of the data path. Therefore the number of multipliers should be as small as possible. Replacing the multiplier completely by shifts and adds will lead to a very high number of operations [201]. As a consequence, a single multiplier is used in the data path. This balances the operation count and the data path complexity. In addition, one shifter and one comparator are selected for the data path. The data path structure is used as the starting point for selecting the structure of the register file. It is also the basis for determining the interconnection network.

The utilization of the functional units depends on the availability of the input operands and on the possibility to store the results of an operation. So far, only the dependencies among the operations where considered as a constraint on the availability of the operands. However, the operands are typically provided by the register file. This means, the register file must have a sufficient number of read ports. But the read ports are also scarce resources. Increasing the number of read ports in the register file will increase the size of the data path and it will slow down the register file. Therefore it is important to select the most suitable number of ports, neither too small (to avoid idle functional units) nor too large (to avoid a complex and slow register file). In the same way the write ports of the register file and the size of the file are crucial as well. The structure of a well suited register file is determined during the register file exploration. As a rule of thumb, the register file should be kept as small as possible for the chosen data path structure. Especially, when the register file is created by logic synthesis from a standard cell library, it becomes a demanding resource (see section 7.9). This means a small performance gain via an increase of the register file is normally not justified by the additional costs, in terms of increasing chip area and increasing delay due to the register file. As an example, Fig. 7.10 shows the performance results for different register file structures when exploring the *mpeg_play* program.

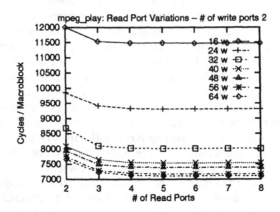

Figure 7.10. Register exploration of the complete *mpeg_play* program. The average over 13 different MPEG sequences is depicted.

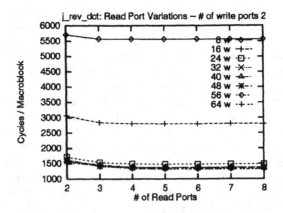

Figure 7.11. Register exploration of the *j_rev_dct* function only.

The graph depicts the number of read ports on the x-axis. The register file size is used as a parameter for the performance curves. The register size is given in terms of 32-*bit* words. The processing time is indicated in cycles per macroblock on the y-axis. The last parameter of the register file is the number of write ports. This parameter is fixed to two write ports for the graphic of Fig. 7.10, however, the HTML framework provides a way to combine several of these graphics into a single HTML page. The links in the page are used to select the appropriate graphic.

The shape of the curves for different register sizes is similar. At least three or four read ports are necessary to support the data path structure that was outlined above. The graphic slightly underestimates the number of ports since an unlimited forwarding among the functional units is assumed (this will be addressed when designing the interconnection network in section 7.11). Hence, a register file structure with four read ports and two write ports is selected. The last parameter of the register file is the size. Small register file sizes like 16 or 24 words show a severe performance penalty. In this case, the data path is not utilized sufficiently. Either the number of functional units in the register file should be decreased or the file must be increased to at least 32 words. In that case, the designer has either the option to increase the size of the register file or to increase the number of ports for gaining additional performance. Increasing the size of the register file beyond 56 words does not give any additional performance gains.

The size also depends on the algorithm that is processed. This becomes apparent by comparing the results of the complete *mpeg_play* program with the results of the *j_rev_dct* function only (Fig 7.11). The *j_rev_dct* function depends on the processing of two vectors, each of them contains 8 words. In addition, constant values are required for scaling operations etc. These values can be stored in approximately 24 words. However, storing more than two vectors does not provide any performance gains because at least 16 of these vectors must be processed (to complete a single one-dimensional IDCT) before intermediate results can be reused again. This observation from the algorithmic properties is confirmed by the results of the exploration. No further performance gain is achieved when increasing the register file size beyond 24 words. This means the performance of the processor is no longer limited by the possibility, to store intermediate results. Again a number of four read ports is sufficient for the data path. Summarizing the results of the register file exploration leads to the following setting of the register parameters. A size of 32 words is the most appropriate, the number of write ports should be at least 2 and the number of read ports should be about 4. However, these port numbers must also be discussed in the context of the interconnection network design. The current exploration step has assumed an unlimited forwarding and a complete connection of all units to all register ports. The next section will address these aspects in more detail.

Data Transfer Analysis

As outlined in section 4.9, there are two parts in the data transfer analysis. The first part is the analysis of the transfer probabilities. This part is mainly influenced by the structure of the program and the algorithmic properties. As with the requirement analysis, little knowledge is necessary about the hardware structure. Only the instruction set is fixed for this part of the data transfer analysis. The second part is the bus exploration. In this case the type of transfer is not considered but the number of transfers. The analysis is closely related to the design exploration. It determines the number of transfers that occur during each clock cycle, for the processor constellation determined in section 7.5. This is another refinement of the exploration to include the information about designing the interconnection network. However, the information is only meaningful for a specific constellation of the processor data path, because the number of transfers depends, e.g., on the number of functional units. Using more functional units will typically require more parallel data transfers, to keep all functional units busy. The bus exploration in this section uses the processor structure that was determined in section 7.5 as the basis of the exploration.

Table 7.6. The data transfer statistics of the PVRG-*p64* program when processing the *stennis.p64* data sequence.

p64: Data transfers

From	Transfers / Macroblock To								
	ADD	*SUB*	*MULT*	*REST*	*COMP*	*SHIFT*	*LOGIC*	*LOAD*	*STORE*
ADD	2383.2	131.6	122.5	13	3029.9	429.8	0.1	2161.9	2182.6
SUB	245.2	151.0	75.5	0	0.2	113.3	188.5	0.0	155.0
MULT	311.8	226.6	8.2	3	0.0	151.0	0.0	0.0	0.2
REST	23.6	0	3	0	10.6	0	0	0	0
COMP	0.0	0.0	0.0	0	0.0	0.0	0.0	0.0	0.0
SHIFT	1179.2	415.3	226.6	0	0.1	11.8	286.5	0.0	457.2
LOGIC	552.6	170.0	11.8	0.2	428.1	178.7	207.7	0.0	174.1
LOAD	1550.8	182.0	318.2	10.3	2758.4	930.5	1065.9	53.1	717.5
STORE	0.0	0.0	0.0	0	0.0	0.0	0.0	0.0	0.0

The description starts with the results of the transfer probabilities. An example thereof is depicted in Table 7.6. The table shows the results for the PVRG-*p64* (*p64* for short) program. The compressed *tennis* sequence (*sten-*

nis.p64) is used as the input stimuli. The complete set of analysis results is available in [59]. The most important data are the transfers between the different operation types. These data are shown in Table 7.6. The transfer data contain the type of operation that produces a certain result (the *from* operation) and those operations which consume this result (the *to* operations). For each of the pairs, the table value shows how often a transfer occurs. A typical example is the incrementing and checking of a loop index variable. In this case an addition is performed to increment the loop variable. Next, a comparison is made to check, if the loop variable has already reached the maximum value. In this case, a transfer is performed from the result of an addition to the input of a compare operation. The same result might be used in a couple of other transfer operations. For example, the result of the addition will be incremented in the next iteration of the loop. In this case it is an additional transfer from an addition to another addition. Table 7.6 contains all these data. Each row of the table corresponds to an operation type that produces a result. The value of in each column is the number of transfers that occur from this producing operation, the row label, to the consuming operation type, that is shown as column header. For example, 131.6 $\frac{transfers}{Macroblock}$ occur from an addition to a subtraction, the second entry in the first row of Table 7.6. In this way the table depicts the communication pattern between the different operation types.

As discussed in section 4.9, it is important to focus on the most important transfers when designing the interconnection network. All other transfers are handled via the register file, i.e., the value is not directly forwarded from one unit to another, but it is stored in the register file and fetched by the consuming unit. While this produces a small performance penalty for the transfer, it improves the overall system performance because the interconnection network can be kept small and fast. It is a different criterion in comparison to the selection of the functional units during the requirement analysis.

In the example of Table 7.6, the most important transfer types are from an addition to a compare operation and from an addition to load / store operations. Other important transfers include the transfer from the load unit to the compare unit. This type of transfer is, for example, important during the decoding, when the values from the coded bit stream are compared against different values in the coding tables. The analysis of the relative importance of transfer types is simplified, if the total number of transfers is considered which starts or ends with a certain operation type. That is, the numbers of the transfers are accumulated along the rows or along the columns of Table 7.6. The results are shown in Fig. 7.12 and Fig. 7.13, respectively. These data can be used to keep the fan-in and fan-out of forwarding connections at a minimum. In this example, the source of a transfer is most often an addition or a load operation. This means the corresponding units are good candidates as sources of forwarding connections. The most important destination types are the comparator and the adder. It is interesting to note that the comparator is an important destination. The design exploration has revealed that the number of comparators can be kept low, i.e., a single comparator is sufficient. However, the comparator is often used. Therefore the performance decreases, if the comparator is not included into the forwarding network. The second most important destination is the adder. This type of transfer occurs for example during an accumulation or when processing the reconstruction. Finally, the load / store units are important destinations due to the operation on data streams in the video compression programs.

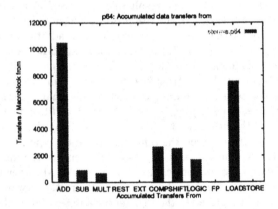

Figure 7.12. Accumulated data transfers from the different operation types (PVRG-*p64* program when processing the *stennis.p64* sequence).

After determining the most significant sources and destinations of transfer operations, it is important to look at the distribution of transfers that occur to and from these important types. The data are shown in Fig. 7.14 and Fig. 7.15, respectively. Fig. 7.14 depicts the number of transfers from an addition to other operations, i.e., the first row of Table 7.6. The most important types in this case are: the transfers of an addition result to a comparator and the use of an addition result as an input to another addition. Furthermore, the transfers from an addition to a load or store operation are significant. This includes the address calculation. An addition is often used to calculate the address of the next operand that should be fetched from memory. The data path determined in section 7.5 contains three adders. This can be used in the interconnection network design. One adder is used for the accumulation. This adder will get a feedback connection from the adder output back to the

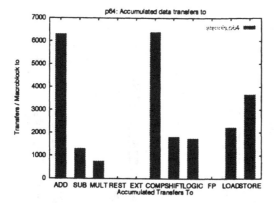

Figure 7.13. Accumulated data transfers to the different operation types (PVRG-*p64* program when processing the *stennis.p64* sequence).

adder input. A second adder can be used for the address calculation and a third adder can be used for the compare operations. This will keep the fan-in and fan-out of the interconnection network small, while still supporting all important connections. Fig. 7.15 shows the most important inputs for the comparator. The figure depicts the $COMP$ column of Table 7.6. The two main inputs for the compare operation are the addition and the load operation. The addition is used when comparing the loop indices against the final value of the loop. The input from the load unit is used, for example, when decoding the variable length data. In this case the coded data in the memory have to be compared against values stored in look up tables. Therefore the main operation is to fetch data from the coded bit stream and to compare these data against the look up table, which is also stored in the memory. This type of operation is reflected in the communication pattern. Thus, the transfer analysis can be used to design the forwarding connections of the processor data path.

So far only a single program and a single data sequence was considered. Table 7.7 extends the results to the investigation of different programs and different data sequences. The table shows the transfers to an addition from other operation types. The addition is one of the most important destinations of transfers in all compression programs. Each column of the table refers to a different data sequence and the corresponding program (*jpeg*, *p64*, *mpeg_play*). Each row refers to a different operation type as source of the transfer operation. The value is the percentage of transfers from the operation (indicated by the row label) to the addition. For example, in case of the *jpeg* program and

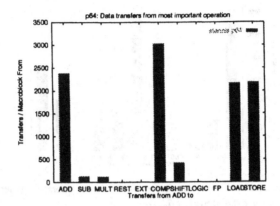

Figure 7.14.   The data transfers from the addition to the different operation types (PVRG-*p64* program when processing the *stennis.p64* sequence).

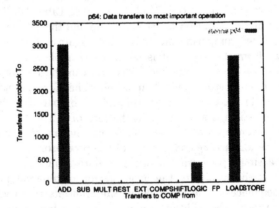

Figure 7.15.   The data transfers to the comparator from the different operation types (PVRG-*p64* program when processing the *stennis.p64* sequence).

the *barb.jpg* sequence (the first column), 42.5% of the transfers to an addition stem from another addition. 23.94% of the transfers to an addition contain the result of a shift operation, etc. The forwarding connections to the adder will concentrate on the most important transfers, i.e., those transfers with the highest percentage values.

Table 7.7.   The percentage of data transfer from different operation types to an addition. The results for different data sequences and different programs are shown.

Data transfer distribution

Transfers to ADD from	Percent of Transfers					
	barb.jpg	lena.jpg	sflowg.p64	stennis.p64	flower.mpg	tennis.mpg
ADD	42.50	43.11	36.52	38.15	51.97	46.07
SUB	3.52	3.58	3.27	3.93	2.73	3.54
MULT	4.54	4.70	5.03	4.99	4.99	7.13
REST	0.00	0.00	0.20	0.38	0.00	0.00
EXT	0.00	0.00	0.00	0.00	0.01	0.01
COMP	0.00	0.00	0.00	0.00	0.00	0.00
SHIFT	23.94	23.65	23.86	18.88	6.74	8.82
LOGIC	1.23	1.16	6.54	8.85	20.32	17.39
FP	0.00	0.00	0.00	0.00	0.00	0.00
LOAD	24.27	23.81	24.59	24.83	13.25	17.04
STORE	0.00	0.00	0.00	0.00	0.00	0.00

Two different cases can be distinguished. First of all, transfers that are important for all the compression programs and all the different data sequences. This is the case for the transfers from an addition to another addition or from the load unit to an addition (i.e., when performing an addition on an operand loaded from the memory). These are good candidates for forwarding connections. The connections cause only a moderate increase of the interconnection network size, but they are useful for all different operating conditions. The second case are transfers like logic to addition. These transfers are not important for the *jpeg* program but they are very significant for the *mpeg_play* program. This means the designer has to balance the negative effect of increasing the interconnection network (more area is occupied and the network gets slower) against the performance penalty that occurs if the transfers are performed via the register file. Another possibility is the use of a multi-functional unit. For example, the logic operations might be allocated to an ALU. The ALU gets a feedback path which allows to use the result of the ALU operation directly as input to the next ALU operation. Since the ALU can perform both, the addition and the logic operation, this allows to implement both transfers: addition to addition and logic to addition. The example shows that the designer has many different options. Therefore the interpretation of the analysis results requires knowledge about these options.

Table 7.8.   The data transfers for the *LoopFilter* function. (PVRG-*p64* program when processing the *stennis.p64* sequence.)

p64: Data transfers of most important function

From	Transfers / Macroblock To								
	ADD	*SUB*	*MULT*	*REST*	*COMP*	*SHIFT*	*LOGIC*	*LOAD*	*STORE*
ADD	558.3	18.3	0.0	0	585.7	109.8	0.0	288.3	421.4
SUB	18.3	0.0	0.0	0	0.0	0.0	36.6	0.0	0.0
MULT	0.0	0.0	0.0	0	0.0	0.0	0.0	0.0	0.0
REST	0	0	0	0	0	0	0	0	0
COMP	0.0	0.0	0.0	0	0.0	0.0	0.0	0.0	0.0
SHIFT	172.0	0.0	0.0	0	0.0	0.0	183.0	0.0	306.2
LOGIC	329.5	18.3	0.0	0	164.7	36.6	54.9	0.0	146.4
LOAD	384.4	18.3	0.0	0	391.2	405.0	366.1	0.0	36.6
STORE	0.0	0.0	0.0	0	0.0	0.0	0.0	0.0	0.0

A last possibility is the implementation of important functions by dedicated hardware. This is often useful if the functions exhibit a regular transfer pattern. In this case, the connections in the hardware can be limited to these particular transfers. The result is a compact realization of the function via hardware. Often filter operations are good candidates. The operations require little controlling and regular operations. The data path can be "hard-wired". Thus, at the first glance, the *LoopFilter* function of the *p64* program seams a good candidate. It requires a high percentage of the operations, in case of the *p64* program. It implements a two dimensional filter. The filter is separable, that is, it can be split into two 1-dimensional filter operations. The filter kernel for the 1-dimensional operation is $\frac{1}{4} \cdot [1\ 2\ 1]$. This is the first deviation from a normal filter. The filter operation can be performed without any multiplications. There is a second special constraint. The filter is applied on an 8×8 block basis. Therefore the filter requires a significant amount of control operations for checking the boundary conditions. In summary, this creates a different situation in comparison to normal filtering. The filtering becomes control dominated and is less regular. This is obvious from the data shown in Table 7.8. A large number of transfers occur from additions and load operations on the one hand to the compare operation on the other hand. Transfers from multiplication to an addition, which are typically considered important for filtering, are not used at all. This example clearly indicates the importance of the analysis. Even the communication pattern of seemingly simple functions, such as the filtering, can become complex due to compiler optimizations such as the strength reduction (i.e., replacing multiplications by shift operations) and specific constrains, such as the checking for boundary conditions. The filtering is no longer a simple data processing operation. Only the analysis can reveal all these specific properties. General assumptions about filtering operations are not sufficient for designing a suitable structure of an interconnection network, in this case.

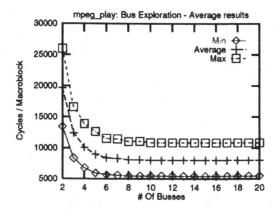

Figure 7.16. The bus exploration data of the *mpeg_play* program. The data for 13 different image sequences are shown.

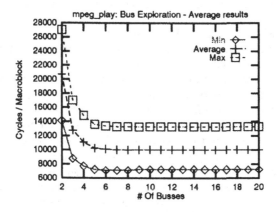

Figure 7.17. The bus exploration data when using a small register file (1 write port, 2 read ports, 32 words).

So far the analysis has focused on the types of transfers. The bus exploration completes these data by investigating the number of transfers that occur concurrently. Fig. 7.16 shows the results of this exploration for the complete *mpeg_play* program. The investigation was performed for the data path structure determined in section 7.5. The processor consists of 3 adders, 2 load units, 1 store unit, 1 multiplier, 1 shifter, and 1 comparator. A register file with 32 words, 4 read ports and 2 write ports is used. The structure of the data path and the register file influences the number of data transfers. The structure determines how operations might be performed in parallel. Using more functional units and a register file with more ports would allow to perform more operations in parallel. Thus, more transfers must be performed. On the other hand, the available degree of instruction level parallelism in the programs bounds the number of parallel operations and therefore the number of required data transfers.

The graph shows the results for 13 different image sequences. Three values are depicted for each number of busses. The minimum and maximum cycle count. These are the values for the sequences requiring the best-case and worst-case processing time, respectively. The bounds show the performance variation when using different image sequences. The average value shows the typical performance that is achieved in the ensemble of image sequences. The results show that approximately 6 transfers occur in each cycle. That is, 6 busses are necessary. Using more than 6 busses does not improve the performance any more, because no more operands are needed in each cycle. The data assume a complete connection of all units to all busses. This means, a bus can be used to transfer a register value or a forwarding result, and each unit can read the value. Thus, the number of busses are slightly underestimated, because the busses are more universal than the busses of a real processor.

The dependence of the bus exploration on the hardware structure of the processor can be seen by looking at a processor structure with a smaller register file (Fig. 7.17). In this case the register file has only 2 read ports and 1 write port. As shown in the design exploration of the register file, it leads to a bottleneck, because the register file cannot provide a sufficient number of operands for the data path. This is reflected by the bus exploration results. The overall processing time for the best-case cycle time increases from $5000 \frac{cycles}{Macroblock}$ to $7000 \frac{cycles}{Macroblock}$. This means, providing more busses does not improve the performance because the operands are missing. The register file is the bottleneck. The second observation is, the saturation of the performance (the knee in the curves) starts earlier. The performance-gains already stop with 4 busses. This means that 3 ports of the register file and one forwarding transfer are used on average per cycle. Further performance improvements are limited by the register file bottleneck.

The two parts of the data transfer analysis have investigated the different properties of the interconnection network. The analysis of the transfer probabilities has investigated the types of transfers which are most important. The bus exploration has shown how many transfers occur in each cycle. This shows, how complex the interconnection network must be for the processor structure determined by the design exploration. The real interconnection network will be designed in the cosynthesis. The cosynthesis combines the two results of the transfer analysis. The connectivity of the functional units to the busses is reduced and the forwarding is limited to the most important transfer types. For example, a direct transfer from an adder result to an adder input will be provided. Connections from the adder and load units to the comparator will be implemented. Finally, the use of multi-functional units will be performed. This helps to achieve better utilization when working under different operating conditions. For example, when *jpeg* and *mpeg_play* lead to different relevance of the transfers from logic results to additions. The next section demonstrates the cosynthesis by looking at the IDCT and different processor structures.

COSYNTHESIS EXAMPLE: IDCT

The cosynthesis implements the specific processor structure. In particular, this includes the design of the interconnection network and the use of multi-functional units. The designer will implement several processors. The basic processor structure is based on the results of the quantitative analysis. The cosynthesis will implement small variations within the vicinity of this basic processor structure. A compiler is generated along with the hardware description of the processor (see section 5.1). This facilitates the software development and it allows to study different algorithmic variations. The variants are implemented as different programs or as versions of certain functions within the programs. This section will exemplify the approach by looking at the IDCT (Inverse Discrete Cosine Transform), as a case study.

The basic definition of the IDCT for the video compression is given in the compression standards [95] [154]. It transforms an 8 × 8 pixel block into a frequency representation. Cosine functions are used for a base system. The IDCT is separable. This allows to split the 2-dimensional IDCT into two 1-dimensional transformations (in x- and y-direction). The one-dimensional transformations can be written as 8 × 8 matrix multiplications (see chapter 4). But there are numerous different alternatives for implementing the IDCT [87] [204]. This makes it an ideal test case for looking at different combinations for hardware, software, and algorithms [204] [118].

The most basic variation of the IDCT stems from exploiting the symmetry properties. In this case the regularity of the data flow in the matrix based implementation is sacrificed. The symmetry of the matrix coefficients is used to reduce the number of operations. The approach is similar to the reduction of operations in a fast Fourier transform [83]. A second way to reduce the operations for the IDCT is by exploiting the knowledge about the compression programs. It takes advantage of the special context where the IDCT is part of a transform coding algorithm. This section discusses five different versions of the IDCT which illustrate the possible variations:

matrix_dct: This is the normal matrix implementation of the IDCT as described in chapter 4. The innermost loop is unrolled and a temporary variable is used for the accumulation of the dot products in the matrix multiplication.

matrix_butterfly_dct: This is the first step towards a fast IDCT. One layer of butterfly operations is used [188]. It allows to replace the 8 × 8 matrix multiplication by two 4 × 4 matrix multiplications. This means some regularity is sacrificed for a reduction of the arithmetic operations.

ChenIDct: This is an implementation of Chen's IDCT [30] which is used in the PVRG programs. The functions are slightly modified to avoid the accumulation without temporary variable (see chapter 4).

j_rev_dct: The fast IDCT which is used in the *mpeg_play* program [149]. The basic algorithm is similar to the *ChenIDct* function. However, more optimizations are applied in the implementation. The implementation is programmed in C. But it is done on a very low level so that many optimizations are done directly via preprocessor macros. Little freedom is left to the compiler. In this way the implementation is close to hand-optimized assembler code.

j_rev_dctOpt: The function which is actually used in the *mpeg_play* program. This implementation takes the basic *j_rev_dct* function and applies additional knowledge about the compression programs. As shown in Fig. 2.2, the IDCT is integrated into the decoding structure. It receives input coefficients from the inverse quantization. The reduction of redundancy in the coder leads to a large number of zero coefficients. It is very likely that the input data for the IDCT are zeros. In this case it is possible to save arithmetic operations. For example, adding a zero input coefficient can be skipped, or multiplying by zero, results in a zero which can be directly propagated to the next operation. The situation is comparable to a constant propagation performed by a compiler. However, these decisions must be performed at run-time. Therefore if-then-else statements are inserted into the code. The statements check for the zero input data. It requires a careful trade-off since the checking needs operations on its own, that is, arithmetic operations are reduced at the cost of additional control operations.

The first step, for evaluating the different variants of the IDCT is a requirement analysis. The results are shown in Table 7.9. The different IDCT functions were integrated into the *mpeg_play* program. The *flower.mpg* sequence was used as input sequence. The second column of the table shows the resulting operations per macroblock. The next column shows the relative operation count in comparison to the *matrix_dct* implementation (the slowest implementation in this case). For example, the *matrix_butterfly_dct* use 71% fewer operations than the *matrix_dct*. The highest improvement is achieved by the *j_rev_dctOpt* function, nearly a factor of 8 in comparison to the *matrix_dct* function. The reduction of the operation count from the *j_rev_dct* function to the *j_rev_dctOpt* function is also significant, a reduction by 50% is achieved. The number of operations spent for the checking of zero coefficients is much smaller than the saved number of operations. Comparing the *ChenIDct* function and the *j_rev_dct* function shows that the manual optimization of the C code (i.e., programming almost on an assembler level) gives also a significant reduction of the operations per macroblock.

The fourth column of the table shows the static number of instructions for the different functions. The matrix based implementations are the most compact implementations in terms of the number of instructions. The fast implementations require an increase in the code length by approximately 30% to 70%. The additional checking for zero coefficients requires more than 4 times as many operations as the *matrix_dct* function. This means the operation count reduction is accompanied by a significant increase in the code size. However, the total code length is still small. It is not beyond the size of the normal instruction cache.

The remaining columns of Table 7.9 show the different instruction mixes. Most interesting in this case is the reduction of the load operations when using the variants which are based on fast algorithms. This is very important because the memory bandwidth is typically a very critical resource. All fast implementations are favorable at that point, because the number of transformation coefficients that are loaded from the memory is smaller. Furthermore, the different characteristics of the *j_rev_dct* function and the *j_rev_dctOpt* function are directly visible by looking at the increase of the compare operations. The amount of compare operations is increased by nearly 10 percent points. This confirms that the improvements were achieved by trading arithmetic operations against a larger percentage of control operations. The next section presents the synthesis results for different processors that will implement these IDCT variants.

Table 7.9. Instruction mix data for different versions of the IDCT. The *flower.mpg* sequence was used as data input.

IDCTs: Instruction mix

Functions				Instruction Mix					
	Op./MB	Improve	Inst.	ALU	MULT	SHIFT	COMP	LD/ST	REST
matrix_dct	15641.1	1	170	42.47	16.36%	4.09%	2.30%	34.78	0.00%
matrix_butterfly_dct	9133.3	1.71	187	43.78	14.00%	8.75%	1.97%	31.5	0.00%
ChenIDct	5520.4	2.83	294	49.72	11.57%	26.40%	0.72%	11.58	0.00%
j_rev_dct	3141.7	4.98	234	50.86	10.36%	9.04%	1.90%	20.3	7.54%
j_rev_dctOpt	2084.7	7.58	869	42.09	7.16%	9.85%	11.69%	19.79	9.43%

Hardware Synthesis

The cosynthesis is subdivided into two parts: the generation of the compiler back-end and the hardware synthesis. The hardware synthesis is similar to a conventional synthesis. The cosynthesis tools (see section 5.1) generate a VHDL description of the processor. The description is based on generic library components [199]. The components allow to configure the processor from large building blocks, rather than developing it from the scratch. The library components are generic. They can be modified by selecting certain parameters, e.g., the size and the number of ports for a register file. This means each of the components in the library represents a complete class of processor building blocks. All processors that are designed via the cosynthesis environment are based on these building blocks. This fact can be used to improve the synthesis speed by incremental synthesis. In this case each of the components is synthesized when it is first used in a processor design. The components are synthesized without taking into account the boundary conditions. They are synthesized as stand-alone components. A synthesized component is stored in a special design library. All subsequent processor designs will use this library for linking. That is, if a processor design needs a component that was already used in a previous design, it will include this component directly from the design library without synthesizing it a second time. This approach sacrifices some optimization possibilities in favor of a faster design time. It is well-suited for exploring many different processor alternatives. Most of the processors will be linked from already existing components. The synthesis tools needs to include only the glue logic for connecting the pre-compiled components. This reduces the time for creating different processors. The final hardware implementation can still use a complete synthesis (including all boundary optimizations), when the best candidate processor is determined. The results for synthesizing the most commonly used components are summarized in Table 7.10 and Table 7.11, respectively. A word width of 32 bit is assumed for all components in the tables.

Table 7.10.   Synthesis results of different library components.

Component Type	Library Element (Parameters)	Gate Count
Register	RegFile (16w, 2in, 4out)	8996
	RegFile (16w, 3in, 6out)	13374
	RegFile (32w, 2in, 4out)	16661
	RegFile (32w, 3in, 6out)	26233
	Forward	352
ICN	MUX (2in)	160
	MUX (4in)	320
	MUX (8in)	1170

Table 7.11.   Synthesis results of arithmetic library components.

Library Element	Gate Count
ALU_ShiftExtend	2524
ALU_Shift	2297
ALU_Extend	1330
ALU	1237
AddSub	1308
Compare	389
Multiplier	17505
ShiftLeftRight	1176

The components are grouped in three types. Table 7.10 shows the results of the register file and multiplexer synthesis. The multiplexers are used primarily in the interconnection network. The third group contains the arithmetic units. It is shown in Table 7.11. All components were synthesized from the LSI10k standard cell library. The Synopsys Design Compiler was used for the synthesis. The register file complexity depends on the size of the register file and on the number of ports. The gate count is approximately linear in the size of the register file. Doubling the size from 16 words to 32 words will also double the gate-count. Furthermore, the gate count is roughly proportional to the number of ports. The register file synthesis is not very efficient because the files are synthesized from flip-flops. A more efficient way of realizing the files will use macro generators or full-custom designs. The gate-count of the multiplexers is also linear in the number of input ports. The delay of the multiplexers is increased significantly when using a large fan-in. This has to be considered when designing the interconnection network.

The arithmetic units are synthesized efficiently. The units were optimized for speed. This shows one weakness of the incremental synthesis approach. The gate-count for the *AddSub* unit is larger than the gate-count of the *ALU*. The *AddSub* unit is slightly faster than the *ALU*, however. In a normal processor, both units should have the same delay, since they are both in the execution stage of the processor. The slowest of the two determines the maximum clock speed. The incremental synthesis treats both unit types individually (i.e., both are optimized for the maximum possible speed). So some hardware for the *AddSub* unit is wasted on an unnecessary speed improvement.

Multi-functional units will typically lead to a gate-count that is approximately proportional to the gate-count of the combination of single functional units. This can be seen when looking at the gate-count of the *ALU_Shift* unit which is approximately in the same order as the gate-count of an *ALU* plus the gate-count of a *ShiftLeftRight* unit. The main advantage of multi-functional units is the saving of connections in the interconnection network. It allows to reduce the overall number of busses and register file ports, as discussed in section 7.11.

The results when synthesizing different processors from the library components are depicted in Table 7.12. The next section will exemplify the codesign possibilities by using these processors for executing the different IDCT variants of Table 7.9. The variation range of the processors in Table 7.12 is larger than typically necessary. It should emphasize the different trade-offs possibilities between algorithmic variations and hardware variations. In most cases the quantitative analysis allows a more precise definition of the principal processor structure before starting the cosynthesis. In this example, the processors range

Table 7.12. Synthesis results for different VLIW processors. The synthesis was performed with the Synopsys Design Compiler using the LSI10k library. Abbreviations: RF: Register File, w: words, in: input ports, out: output ports.

Processor synthesis results

Processor	Architectural	Gate Count				
		Regs	*Units*	*Connect*	*Rest*	*Total*
P_A	RF(16 w, 2 in, 4 out), 1 ALU_Shift, 1 Compare, 1 MPY	8996	20418	2080	5	31499
P_B	RF(16 w, 3 in, 6 out), 2 ALUs, 1 Compare, 1 MPY, 1 Shift,1 Forwarding	13726	21637	4050	13	39426
P_C	RF(32 w, 1 in, 2 out), 2 ALUs, 1 Compare, 1 MPY, 1 Shift, 4 Forwarding	12277	21730	6870	0	40877
P_D	RF(32 w, 2 in, 4 out), 2 ALUs, 1 Compare, 1 MPY, 1 Shift	16661	21637	4050	43	42391
P_E	RF(32 w, 3 in, 6 out), 2 ALUs, 1 ALU_Shift, 1 Compare, 1 MPY, 1 Shift, 5Forwarding	27993	23934	5170	15	57112

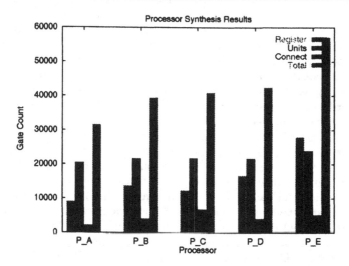

Figure 7.18.   The gate count of different processors from Table 7.12.

from an almost scalar processor P_A to processor P_E which uses a high degree of instruction-level parallelism.

Each of the five processors in Table 7.12 uses a comparator and a multiplier. The comparator is uncritical in terms of the gate count complexity, whereas the multiplier is an extremely demanding arithmetic resource (Table 7.11). But investigations in [201] show that replacing the multiplications by shift and add operations will lead to an unacceptably high number of functional units, for all but the smallest image formats. Thus, it is necessary to have a multiplier for the video compression programs, although it increases the gate count of the functional units significantly. The number of remaining arithmetic units is increased from processor P_A to processor P_E. Processor P_A uses only a single *ALU_Shift* unit. This means, only one arithmetic operation will be performed in each cycle. On the other hand, processor P_E has 3 *ALUs* and an additional shifter. Up to four arithmetic operations can be performed in addition to a multiplication and a compare operation. Additionally, the interconnection network is also extended. Processor P_E can use up to five forwarding connections which allow a bypassing of the register file.

The second major variation of the processors is along the register file. Different sizes and different numbers of ports are used. Processor P_A and processor P_B use a small register file with only 16 words. The design exploration of section 7.5 already indicated, that such a small register file will reduce the performance, but, as shown in Table 7.10, the register file is also a demanding component in terms of the gate-count. Therefore it is important to analyze the effect of using a smaller register file. The remaining processors use register files of 32 words. The number of ports is changed for each of the different register sizes. Processor P_B and processor P_E use a register file with a total of 9 ports. This means, a maximum of 6 operands can be read out of the register file in each cycle. The file can store 3 results at the same time. Besides this, the forwarding of the results among the functional units can be used to bypass the register file altogether.

A comparison of the gate-count for the different parts of the data path is depicted in Fig. 7.18. The comparison shows that the gate-count of the functional units is dominated by the gate-count of the multiplier. The larger number of functional units in processor P_E leads only to a small relative increase of the total gate-count. The increase of the gate-count due to the different register structures is more significant. The complexity of the register file can even exceed the complexity of the functional units. The register gate-count is nearly tripled when going from processor P_A to processor P_E. But the large register file for processor P_E is necessary. The larger number of functional units can only be supported, if the data path is accompanied by a more complex register file, which can provide the required number of operands for the functional units. This leads to an overall increase of the gate-count from 37k Gates to 57k Gates for the data path of processor P_E.

The gate-count of the interconnection network is comparatively small. But the higher wiring overhead has to be taken into account. The larger interconnection network will have a higher delay, which means that the clock speed of the processor must be reduced. The next section will investigate the performance of the different IDCT variants on the five processors.

Codesign

The second part of the cosynthesis is the generation of a compiler along with the processor. This provides a way to create software for many different processor variants without having to develop assembler code for each of the processor variants. As an example, the different variants of the IDCT from section 7.17 will be used. Each of the variants is compiled for the five processors of section 7.9. The results of an instruction-set simulation for the *flower.mpg* sequence are shown in Table 7.13. Such an investigation would be next to impossible if performed in a conventional design style. It would mean to develop 25 different assembler programs just to test the best combination of hardware and software. In particular, if the processors are not available, this is a tedious work. The software development must be done on a simulation model of the processors. Typically, this is a VHDL model on the RT level. It leads to extremely large run-times for the debugging.

The cosynthesis environment achieves a significant reduction of the design time at this point. It allows to develop variants of the IDCT as C programs. The variants can be linked into the normal *mpeg_play* program, which was developed for a workstation or PC. The new versions can be tested by directly executing them. The normal debugging environment can be used. This means, the development of the different variants is much faster, because the functional correctness can be tested in the designer's normal programming environment. It is even possible to use small image sequences for testing the image quality by inspection (although a normal workstation might not achieve real-time speed). In this way, the software development, when using the cosynthesis environment, is as powerful as a normal software development. Furthermore, the number of manually developed variants is reduced. The designer develops only the five different algorithmic variants. The implementation of these variants on the processors is done automatically via compiling the programs with the generated compiler back-end. This gives the foundation for exploring the combined design of hardware, software, and algorithms.

Table 7.13 shows the cycles per macroblock when using the different IDCT variants. The comparison of the improvements is facilitated by normalizing the

Table 7.13. Processing times per macroblock for executing the different IDCT functions on the five different processors. The *flower.mpg* sequence was used as data input.

IDCTs: Processing Times

Function	Cycles / Macroblock				
	P_A	P_B	P_C	P_D	P_E
matrix_dct	24011.49	24235.90	17301.74	13329.74	8639.65
matrix_butterfly_dct	15573.81	15551.37	9963.65	7450.29	5183.79
ChenIDct	7450.29	6058.97	5295.99	3882.23	2715.32
j_rev_dct	5969.21	4847.18	3141.69	2226.11	1463.13
j_rev_dctOpt	4510.57	3949.55	2490.91	1851.35	1577.58

processing times. This is shown in Fig. 7.19. The processing time is normalized w.r.t. the slowest implementation. In this case it is the *matrix_dct* function and processor P_B. The speed up in comparison to this combination is shown in Fig. 7.19.

First of all, the algorithmic improvements can be examined for a single processor, such as processor P_A. In this case the *matrix_dct* function is the slowest implementation. The *j_rev_dctOpt* function achieves approximately a speed up factor of 5. This factor is slightly smaller than the improvements obtained in terms of the total operation count (Table 7.9). In particular, the improvements due to the checking for zero data coefficients are not as efficient as suggested by the reduction of the operation count. The *j_rev_dct* function achieves a speed up of 4 which is comparable to the reduction of operations. But the *j_rev_dctOpt* function achieves only a reduction by 5 whereas the reduction of the operation count was 7.5. This shows, that even for the small processor P_A the hazards in the pipeline, due to the increased number of branches, are significant. The other variants achieve approximately the improvement suggested by the reduction of the operation count.

The second axis of variation is along the different hardware structures of the processors. Processor P_A is an almost scalar processor which uses little instruction-level parallelism. On the other hand, processor P_E can perform several operations in parallel. Comparing the improvements for a single algorithm, such as the matrix multiplication, leads to a total speed up of approximately 3. The small register files of processor P_A and processor P_B produce a severe performance bottleneck. Even the increase of the number of ports in processor P_B gives only a modest performance improvement in return. Furthermore, the larger number of functional units in processor P_B cannot be used efficiently. The performance of processor P_D is significantly better, due to the larger register size. The second improvement of the register file is the increase in the number of ports. This also gives a better performance gain when using it in conjunction with the larger register file. This can be seen by comparing the improvements of processor P_C and processor P_E. The adaption of the register file and the data path is very crucial for gaining performance from a larger number of functional units or from using more forwarding connections. This adaptation is mainly done during the register file exploration.

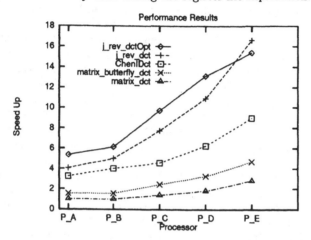

Figure 7.19.   Performance of the different IDCT functions (Table 7.9) on the five different processors (Table 7.12). The processing time is normalized w.r.t. the slowest implementation.

The most important result from the codesign point of view is the combined speed up when looking at the processor variation and the algorithmic variation at the same time. This means the codesign allows to select the combination of algorithms and hardware, whereas the classical design would select first the algorithm and then the hardware structure, or vice versa. The classical approach leads not to the best combination: the best algorithm, in terms of the operation count, is the *j_rev_dctOpt* function. The processor with the best performance is the processor P_E. Combining both selection gives *not* the best overall result. In other words, the classical design approach optimizes individually along each axis of the design domain (see Fig. 2.3). The codesign achieves a better result because it allows to optimize in the complete design space. The smallest processing time is achieved when combining the processor P_E and the *j_rev_dct* function.

Processor P_E cannot benefit from the improvements in the *j_rev_dctOpt* function because the control unit becomes a bottleneck. The improvements of the *j_rev_dctOpt* function were made by saving arithmetic operations at the cost of additional branches. This is costly for a processor with a high degree of instruction-level parallelism. In this case the processor has enough functional units to process several arithmetic operation concurrently, but the branches will introduce pipeline stalls which lead to idle cycles in the processor. It is more efficient to use an algorithm like the *j_rev_dct* function where a higher number of arithmetic operations is required (this can be provided, due to the high degree of instruction-level parallelism) but less branches.

But the embedded system design is even more complex. It is not always desirable to design the fastest possible solution, rather, the most cost efficient for the specific design context is of interest. At this point the real-time condition of section 2 must be considered. For example, if a hand-held video phone is the design target, this will typically be used in conjunction with a slow channel. Therefore only small image formats need to be considered. A small processor like processor *P_A* is sufficient to fulfill the real-time condition. In this case the best combination is processor *P_A* and the *j_rev_dctOpt* function. This means there is no single best algorithm or processor. It is always the combination of algorithms, software, and hardware which has to be considered. This combination has to be optimized for the specific conditions of the embedded system design. The designer can take advantage of the operating conditions, such as the real-time condition. It gives him/her the possibility to find cost efficient solutions for particular design conditions. The crucial elements in the design process are the quantitative analysis and the cosynthesis. They allow to investigate the different trade-offs without using excessive design times.

LESSONS LEARNED AND FUTURE WORK

Designing embedded systems is a complex task because the work from many different disciplines must be combined. This makes the system design itself inherently complex. Developing software for the design automation suffers from the same problems as many other large software project [58]. Therefore it is important, to summarize the experiences gained from developing the CASTLE tools described in the thesis. First of all, it was very important to have a real application example. The *design automation* has to start with a design. It provides invaluable feedback for designing the tools. In particular, it avoids to create automatic algorithms which turn out to be useless in practice. The design tools should gradually evolve according to the experiences gained from the design. This means, the development should start with a small set of tools. The tools are tested by applying them to a design project. Therefore it is necessary to have working prototypes of the tools [184]. A first prototype of the analysis environment [201] influenced the design of the final system significantly. For example, two important features were included into the analysis environment after testing the prototype: (1) the necessity for including a shell interface and (2) the support for an automatic analysis. Both features were developed after performing a number of measurements manually. It is important for this evolutionary development to have small tools. They are later combined into larger tools [206]. In this way the system can be constructed in a stepwise manner. Each tool can be used on its own. This is extensively used during the development of the design exploration environment. In particular, the simple interpreters allow to integrate the individual modules into a variety of other languages such as Tcl or the EMACS systems. But all these modules can be used on their own and they might be replaced by other, more advanced versions, in the future. This facilitates the extension of the complete system. The modules can be adapted if certain inconveniences are discovered in the practical application.

The main advantage of this approach is the steady progress in the tool development. The costs for this controlled development are the increased number of small interpreters and interfaces among the autonomous tools. These costs are small in comparison to the costs that occur if a large and ambitious software system must be changed at a late stage of the design, or if it turns out to be inconvenient for the practical use.

Besides designing new modules for specific tasks, it is also important to integrate existing software as much as possible. Re-developing existing software should only be done in very exceptional cases. The integration of existing software turned out especially powerful in the HTML-based codesign framework. In this case the EMACS system and conventional WWW browsers are used as the main interface for the system. This increases the familiarity for the user. For example, the EMACS possibilities, like command completion, sophisticated help, customization by adding modes, etc., can all be used directly. Furthermore, it improves the reliability due to the larger user community. Finally, the approach profits from the gains made in other fields. This is especially important for the networking, which is rapidly evolving. The user of the HTML-based framework can directly profit from all these tools. This will be a major factor for future developments. Computers will rely more and more on the integration into the network. The network will provide most of the infrastructure that is necessary to use a computer. Furthermore, the design of complex embedded systems will require expertise in many different fields. This is often only available, by building design groups that are distributed at many different sites. This will require the frequent exchange of design data and ideas via computer networks. Therefore it is important to integrate the design tools directly into this world of network computer systems.

Two major aspects have to be addressed in the network context: the distribution of the analysis data and the filtering of information from large databases. Part of this is already done in the HTML-based design framework. The data are formatted and integrated in such a way, that they can be accessed via the WWW. No special design tools are required for the access. The measurement and the analysis of the data is split. This is extremely important, because it allows to shift the analysis from the server side to the client side of the network. The designer will use the pre-calculated analysis data in the HTML pages to get an overview of the analysis. Next, the designer will start network based analysis tools. An example would be a tool written in a language, like JAVA [185], which is integrated into a conventional HTML browser. The tool can be distributed along with the fine-grained analysis data. The tool will install itself on the client side. It allows to analyze the data in many different perspectives. All this can be done without installing design tools in advance. Only the programs for the current analysis are distributed. This gives the designer the additional advantage of having his/her own perspective on the analysis data while still sharing the costs for performing the measurements within the design group. Each designer can use the analysis data in different ways.

This emphasizes a second problem: the information filtering. It will become an essential feature for using the automatic quantitative analysis. Due to the automation, it is possible to analyze many different variants of the system structure. An example is the variation of algorithms on many different processor structures. All this can be measured automatically. While the individual measurement is very time consuming, it can be done in parallel on many different computers in a distributed computer network. This means large amounts of data are created in comparatively short time. As a consequence, the designer has to sort out the "relevant" information from the large set of information. The relevance might be different for the designers. For example, a software engineer might like to know how to improve the program implementation, whereas the hardware engineer might focus on the changes in the hardware structure that are required when using certain program implementations. Extracting the different information from the large number measured data will reveal much information about the design. It allows the designer to perform design decisions more efficiently. Therefore it is a key component for improving the overall design efficiency. It will also improve the transmission of the analysis data, because the filters allow to extract the important information while sorting out the unimportant data. This saves transmission bandwidth and reduces the costs for cooperating via networks.

The architecture extension is going to be a second goal of future work. This will raise the system performance beyond the limits imposed by the instruction-level parallelism. There are two major directions of improvements: (1) specialized functional units and (2) heterogeneous multiprocessor systems.

Specialized functional units offer the possibility to support complex instructions directly in hardware. Examples are the multimedia extensions of standard processors (see section 1.1) or array processors for blockmatching (see section 2.4). The specialized functional units are adapted to specific types of tasks, e.g., parallel operation on small integers in case of the multimedia extensions. The specialization accomplishes an efficient utilization of the hardware. Thus providing high performance at comparatively low costs. Specialized units can be directly integrated into the VLIW concept. From the hardware point of view, they are integrated as additional components into the library. They are implemented by high-level synthesis or full-custom design.

The major problem is the compiler support. A retargetable compiler must "know" the functionality of the specialized units. The translation of the high-level language (e.g., C/C++) must be done in such a way that the special instructions are exploited. This is often very difficult. The designer can improve the efficiency by writing the program in a specific way or by guiding the compilation with directives. These directives will annotate the program. The quantitative analysis will be used to focus these manual optimizations on the important parts of the program.

The efficient utilization of specialized units gets more difficult if the units get more complex. Often, complex instructions have a very limited application range which makes them difficult to support in a compiler [76]. Heterogeneous multiprocessor systems offer an alternative at that point. They provide the opportunity to partition the tasks of the application program on a coarse grained level. Several processors are combined which are adapted to different types of tasks. An example is a standard superscalar processor in combination with a VLIW processor. The resulting system might be used for video compression. The standard microprocessor would be used for tasks like the Huffman decoding which offer little parallelism. The VLIW processor can focus on tasks like the IDCT which are more amenable for a higher degree of instruction-level parallelism (see section 7.5).

The main problem in this case is the efficient communication and synchronization of the process. For example, a shared memory system might be used. Often it will take different memory banks to support the parallel data access. This requires a data layout which allows a conflict free access. In embedded system design this must be done with little overhead. In the same way the synchronization among the processors is crucial. Again an efficient mechanism is required. This may be specific for a certain application domain. For example, in video processing a macro-pipelining of tasks is useful [202]. The designer will need analysis tools for estimating the communication overhead. In addition, a support is required for inserting the synchronization and for performing the data layout.

The combination of both approaches (heterogeneous multiprocessor systems and specialized functional units within the processors) offers a significant potential for performance improvements especially when looking at embedded system. The narrow application domain permits many ways of exploiting specializations.

8 CONCLUSIONS

A design approach for flexible and programmable embedded systems was presented. The design is targeted at a narrow application domain. This thesis has focused on the real-time video compression domain. The most essential point of the methodology is the combined design of hardware, software, and algorithms. It is achieved in two main steps: the *quantitative analysis* and the *cosynthesis*. The quantitative analysis provides a picture of possible design trade-offs without implementing the system in detail. The designer uses this picture to develop the principal system structure. This is done on an abstract level. Next, the cosynthesis translates the abstract system structure into the final hardware and software implementation. This is a designer centered approach. It combines design quality (by using the designer's knowledge) and design efficiency (by using tools for low-level tasks).

The approach was applied for the design of an embedded video compression system. A paradigm for the documentation and the integration of the design tools was developed. It is based on the WWW (World-Wide Web), thus facilitating the design work in environments with computer networks. A set of design tools was built. It supports the designer centered approach. This required new features from the design tools. A visualization of the result data becomes cru-

167

cial. It must be combined with a fast access to the most important parts of the design information. In other words, the tools act as information filters. This is essential for system-level design. The tools can create large amounts of data. But the designer can handle only a limited complexity. This means, the tools must extract the design information which gives the best returns when the designer applies his/her time and context knowledge.

An example thereof is the quantitative analysis. It allows, for example, to determine the most important functions of a complex program. The results of the video compression programs show that only a small percentage of the functions is performance critical. The designer can focus on these important functions when improving the design. The quantitative analysis opens the possibility to concentrate the optimizations. The costs are no longer depended on the complete optimization of a complex program, rather, the design efficiency is improved by focusing on parts of the program where the gain of the invested design time is highest.

Accessing the important design information is further simplified by integrating the results into the WWW. Hypertext browsers can be used to navigate on the design data. Background information can be included into the result data. This is important in embedded system design because of its multi-disciplinary character. Often, designers with different expertise work with the same data. A hypertext framework facilitates this cooperation because team members with less specific knowledge have access to the background information. An additional advantage of the WWW is the distribution of design data on a world-wide scale. This allows to form design groups with experts from different locations.

Another important feature of the designer centered approach is the use of templates. Creating a template requires knowledge and design time. But this time is spend only once. The template can be reused in every subsequent design. It is adapted to the new context by changing a small number of parameters. Again this combines designer knowledge (for creating the template) with automation (when copying and adapting the template). Examples include the VLIW processor template in the schematic entry. The designer can specify a complete processor by defining a small number of parameters (e.g., the number and type of functional units, the size of the register file, etc.). The schematic entry automates the implementation details due to the VLIW processor template. The design is like filling out a form for a processor.

In the same way the generic component library allows to customize the processor components. Again, design time is saved because components are created only once. They are reused in every processor design. Even the speed of the synthesis can be improved by using an incremental synthesis approach. Each component is only synthesized when it is first encountered in a processor design. This is especially powerful when exploring variants of a basic processor structure during the cosynthesis.

Another point of attention in the tool design was the integration of existing tools as far as possible. In particular, this was used in the HTML environment. The top-level interface is done via the EMACS system. Conventional HTML browsers, like Netscape or Mosaic, can be used for accessing the results. Existing tools reduce the learning time for the designer. Additionally, the design tools can take advantage from the improvements made for these tools. The design tools profit from the larger user community.

The use of standards is a second way to improve the coupling of design tools. Examples are: the HTML framework for the results or the storing of result data as ASCII tables. These tables can be used as inputs to spread sheets or they are use as inputs for text processing systems when creating additional design documentations. Furthermore, they open the possibility to render the results in many different ways. Interfaces like this are important to make the design system scalable. In particular, in system-level design this is a major advantage because a large portion of the design time is devoted to documentation or design management. Integrating these tasks into the design system increases the design efficiency.

The design methodology was applied to the design of an embedded video compression system. The quantitative analysis was carried out for the three major video compression standards (JPEG, H.261, MPEG). An ensemble of typical input video sequences was used. Three major analysis steps were performed: (1) the requirement analysis, (2) the design exploration, and (3) the data transfer analysis. The requirement analysis is mainly concerned with the investigation of the algorithmic properties and the software implementation. The design exploration investigates the performance on variations of the processor hardware, but the interconnection network is excluded. The design of the interconnection network is addressed in the data transfer analysis where the most important connections among the functional units are determined. All results are integrated into the WWW [59]. The data are used for a coarse definition of the processor structure. This is the entry point for the cosynthesis. It gives the designer the unique possibility to explore many different combinations of program variants on several processor structures. This would be an extremely time-consuming task in a conventional design process. New opcode must be developed for every combination of processor and program. The cosynthesis simplifies this task. Program variants are developed only once. The compiler is used to create the opcode of the different processors. This was exemplified by a case study of the IDCT. It is one of the most important tasks in all three standards. Variations of it were integrated into the *mpeg_play* program (the fastest compression program). The variations use different algorithms and different software implementations. A number of different processors were developed and the combinations of hardware, software, and algorithms were investigated. The results demonstrate that only the combined optimization achieves the best overall performance in embedded system design.

References

[1] B. Ackland. The role of VLSI in multimedia. *IEEE J. Solid-State Circuits*, 29(4):381–388, April 1994.

[2] B. D. Ackland et al. A video-codec chip set for multimedia applications. *AT&T Technical Journal*, 72(1):50–66, January/February 1993.

[3] N. Ahmed, T. Natarajan, and K. R. Rao. Discrete cosine transform. *Trans. IEEE Computers*, 23(1):90–93, January 1974.

[4] A. Alomary et al. PEAS-I: A hardware/software co-design system for ASIPs. In *Euro DAC*, pages 2–7, Hamburg, 1993.

[5] S. Antoniazzi, A. Balboni, W. Fornaciari, and D. Sciuto. HW/SW codesign for embedded telecom systems. In *Int. Conf. Computer Design*, pages 278–281, Cambridge, MA, Oct. 10-12, October 1994.

[6] M. Antonini et al. Image coding using wavelet transform. *IEEE Trans. Image Processing*, 1(2):205–220, April 1992.

[7] K. Aono et al. A video digital signal processor with a vector-pipeline architecture. *IEEE J. Solid-State Circuits*, 27(12):1886–1894, December 1992.

[8] R. Aravind et al. Image and video coding standards. *AT&T Technical Journal*, 72(1):67–89, January/February 1993.

[9] B. Arnold. Mega-issues plague core usage. In *ASIC & EDA*, pages 46–58, April 1994.

[10] M. Auguin, F. Boeri, and C. Carriere. Automatic exploration of VLIW processor architectures from a designer's experience based specification.

171

In *3rd Int. Workshop on Hardware/Software Codesign*, pages 108–115, Sept. 22-23, Grenoble, September 1994.

[11] M. Auguin et al. Towards a multi-formalism framework for architectural synthesis: The ASAR project. In *3rd Int. Workshop on Hardware/Software Codesign*, pages 25–32, Sept. 22-23, Grenoble, September 1994.

[12] D. Auld et al. A flexible chip set for intra frame video compression. In *Compcon Spring'91*, pages 330–332, 1991.

[13] D. Bailey. The virtues of programmability. In *Computer Design*, pages 71–76, July 1993.

[14] D. Bailey et al. Programmable vision processor/controller. In *IEEE Micro*, pages 33–39, October 1992.

[15] W. Baker. An application of a synchronous/reactive semantics to the VHDL language. Report UCB:ERL-93-10, UC at Berkeley, 1993.

[16] M.F. Barnsley and L.P Hurd. *Fractal Image Compression*. AK Peters Ltd., Wellesley, MA, 1993.

[17] F. Baskett and J.L. Hennessy. Microprocessors: From desktops to super-computers. *Science*, 261:864–871, 13August 1993.

[18] A. Benveniste and G. Berry. The synchronous approach to reactive and real-time systems. *Proc. IEEE*, 79(9):1270–1282, September 1991.

[19] T. Berners-Lee et al. The world-wide web. *Comm. ACM*, 37(8):76–82, August 1994.

[20] M. Bierling. Displacement estimation by hierarchical blockmatching. *Proc SPIE Visual Communication & Imake Processing*, 1001:942–945, 1988.

[21] W. P. Birmingham, A. P. Gupta, and D. P. Siewiorek. MICON: Automated design of computer systems. In Camposano and Wolf [28], pages 205–229.

[22] M. Bolton et al. A complete single-chip implementation of the JPEG image compression standard. In *IEEE Custom Integrated Circuits Conference*, pages 12.2.1–12.2.4, 1991.

[23] N. S. Borenstein. MIME: A portable and robust multimedia format for internet mail. *Multimedia Systems*, 1(1):29–36, 1993.

[24] F. Boussinot and R. De Simone. The ESTEREL language. *Proc. IEEE*, 79(9):1293–1304, September 1991.

[25] K. Buchenrieder et al. HW/SW co-design with PRAMs using CODES. In D. Agnew et al., editors, *Computer Hardware Description Languages*. IFIP Trans., vol. A-32, 1993.

[26] J. T. Buck, S. Ha, E. A. Lee, and D. G. Messerschmitt. Ptolemy: A framework for simulating and prototyping heterogeneous systems. *Int. Journal of Computer Simulation*, 4:155–182, April 1994.

[27] R. Camposano and J. Wilberg. Embedded system design. *J. Design Automation for Embedded Systems*, 1(1/2):5–50, 1996.

[28] R. Camposano and W. Wolf, editors. *High-Level VLSI Synthesis*. Kluwer Academic Publishers, Boston, 1991.

[29] R. Chandra, A. Gupta, and J.L. Hennessy. COOL: An object-based language for parallel programming. *IEEE Computer*, 27(8):13–26, August 1994.

[30] W.H. Chen, C.H. Smith, and S.C. Fralick. A fast computational algorithm for the discrete cosine transform. *IEEE Trans. Commun.*, 25:1004–1009, 1977.

[31] J. Cocke and V. Markstein. The evolution of RISC technology at IBM. *IBM J. Research and Development*, 34(1):4–11, January 1990.

[32] D. Conner. ESDA tools. In *EDN*, pages 80–91, May 26, May 1994.

[33] T. H. Cormen, C. E. Leiserson, and R. L. Rivest. *Introduction to Algorithms*. MIT Press, Cambridge, MA, 1990.

[34] H. Corporaal and J. Hoogerbrugge. Code generation for transport triggered architectures. In P. Marwedel and G. Goossens, editors, *Code Generation for Embedded Processors*, pages 240–259. Kluwer Academic Publishers, 1995.

[35] H. Corporaal and H.J.M. Mulder. MOVE: A framework for high-performance processor design. In *Proc. Supercomputing '91*, Albuquerque, November 1991.

[36] P.C. Cosman et al. Using vector quantization for image processing. *Proc. IEEE*, 81(9):1326–1341, September 1993.

[37] Plessey data sheet. VP 2611. March 1992.

[38] Plessey data sheet. VP 2615. March 1992.

[39] I. Daubechies. *Ten Lectures on Wavelets*. SIAM, Philadelphia, 1992.

[40] G. DeMicheli. *Synthesis and Optimization of Digital Circuits*. McGraw-Hill, New York, 1994.

[41] Ed F. Deprettere. Example of combined algorithm development and architecture design. *Integration, the VLSI J.*, 16(3):199–220, Dec 1993.

[42] R.A. DeVore, B. Jawerth, and B.J. Lucier. Image compression trough wavelet transform coding. *IEEE Trans. Information Theory*, 38(2):719–746, March 1992.

[43] K. Diefendorff and M. Allen. Organization of the motorola 88110 RISC microprocessor. *IEEE Micro*, 12(2):40–63, April 1992.

[44] K. Diefendorff, R. Oehler, and R. Hochsprung. Evolution of the powerPC architecture. *IEEE Micro*, 14(2):34–49, April 1994.

[45] K. Diefendorff and E. Silha. The powerPC user instruction set architecture. *IEEE Micro*, 14(5):30–41, October 1994.

[46] T. Ebrahimi, E. Reusens, and W. Li. New trends in very low bitrate video coding. *IEEE Proc.*, 83(6):877–891, June 1995.

[47] R. Ernst, J. Henkel, and Th. Benner. HW/SW cosynthesis for microcontrollers. In *IEEE Design & Test*, pages 64–75, December 1993.

[48] M.H. Ahmad Fadzil and T.J. Dennis. Video subband VQ coding at 64 kbit/s using short-kernel filter banks with an improved motion estimation technique. *Signal Processing: Image Communication*, 3:3–21, 1991.

[49] C. P. Feigel. TI introduces four-processor DSP. Report CA 95472, newsletter, March 28, March 1994.

[50] R. L. Fetterman and S. K. Gupta. *Mainstream Multimedia: Applying Multimedia in Business*. Van Nostrad Reinhold, NY, 1993.

[51] G. C. Fox and W. Furmanski. The physical structure of concurrent problems and concurrent computers. In R.J. Elliott and C.A.R. Hoare, editors, *Scientific Applications of Multiprocessors*. Prentice Hall, 1989.

[52] H. Fujiwara et al. An all-ASIC implementation of a low bit-rate video codec. *IEEE Trans. Circuits And Systems On Video Technology*, 2(2):123–134, June 1992.

[53] K. Gaedke et al. A VLSI based MIMD architecture of a multiprocessor system for real-time video processing applications. *J. of VLSI Signal Processing*, 5:159–169, 1993.

[54] D. Gajski, N. Dutt, A. Wu, and S. Lin. *High-Level Synthesis: Introduction to Chip and System Design.* Kluwer Academic Publishers, Norwell, MA, 1992.

[55] D.J. Le Gall. The MPEG video compression algorithm. *Signal Processing: Image Communication*, 4:129–140, 1992.

[56] A. Gersho. Advances in speech and audio compression. *Proc. IEEE*, 82(6):900–918, June 1994.

[57] A. Gersho and R. M. Gray. *Vector Quantization and Signal Compression.* Kluwer Academic Publishers, Dordrecht, Netherlands, 1992.

[58] W. W. Gibbs. Software' s chronic crisis. *Scientific American*, 271(3):72–81, September 1994.

[59] GMD.   The   CASTLE   analysis   environment.   http:// alcatraz.gmd.de:9422/ designenv/ hello.html.

[60] J. Gong, D.D. Gajski, and A. Nicolau. A performance evaluator for parameterized ASIC architectures. In *Proc. EuroDAC*, 1994.

[61] R.J. Gove. Architecture for single-chip image computing. *Proc SPIE Image Processing and Interchange*, 1659:30–40, 1992.

[62] R. M. Gray. *Entropy and Information Theory.* Springer-Verlag, New York, USA, 1990.

[63] R.M. Gray. Vector quantization. *IEEE ASSP Magazine*, 1(2):4–29, April 1984.

[64] A.P. Gupta, W.P. Birmingham, and D.P. Siewiorek. Automating the design of computer systems. *IEEE Trans CAD*, 12(4):473–487, April 1993.

[65] P. Gupta et al. Experience with image compression chip design using unified system construction tools. In *DAC*, pages 250–256, 1994.

[66] R.K. Gupta, C.N. Coelho, and G. De Micheli. Program implementation schemes for hardware-software systems. In *IEEE Computer*, pages 48–55, January 1994.

[67] R.K. Gupta and G. De Micheli. Hardware-software cosynthesis for digital systems. *IEEE Design & Test*, 10(3):29–41, September 1993.

[68] K. Guttag et al. A single-chip multiprocessor for multimedia: The MVP. In *IEEE Computer Graphics & Applications*, pages 53–64, November 1992.

[69] K. M. Guttag. Multimedia powerhouse. In *Byte*, pages 57–64, June 1994.

[70] N. Halbwachs. *Synchronous Programming of Reactive Systems*. Kluwer Academic Publishers, 1993.

[71] N. Halbwachs, P.Caspi, P. Raymond, and D. Pilaud. The synchronous dataflow programming language LUSTRE. *Proc. IEEE*, 79(9):1305–1320, September 1991.

[72] T.R. Halfhill. AMD vs. superman. *BYTE*, 19(11):95–101, November 1994.

[73] D. Harel. Statecharts: A visual formalism for complex systems. *Sience of Comp. Programming*, 8(3):231–275, 1987.

[74] B.S. Haroun and M.I. Elmasry. Architectural synthesis for DSP silicon compilers. *IEEE Trans CAD*, 8(4):431–447, April 1989.

[75] J. L. Hennessy et al. Hardware/software tradeoffs for increased performance. In *Proc. Symp. Architectural Support for Programming Languages and Operating Systems*, pages 2–11, 1983.

[76] J.L. Hennessy and D. A. Patterson. *Computer Architecture: A Quantitative Approach*. Morgan Kaufmann Publ., 1990.

[77] S. Hiranandani, K. Kennedy, and C.-W. Tseng. Compiling fortran d for MIMD distributed-memory machines. *Comm. ACM*, 35(8):66–80, August 1992.

[78] F. Hoeg, N. Mellergaard, and J. Staunstrup. The priority queue as an example of hardware/software codesign. In *3rd Int. Workshop on Hardware/Software Codesign*, pages 81–88, Sept. 22-23, Grenoble, September 1994.

[79] S. Hoffos et al. *CD-I Designers Guide*. McGraw Hill, London, 1992.

[80] J. Hoogerbrugge. *Code Generation for Transport Triggered Architectures*. PhD thesis, Delft Technical University, Delft, Netherlands, 1996.

[81] J. Hoogerbrugge and H. Corporaal. Transport-triggering vs. operation-triggering. In *Proc. ACM Int. Conf. Compiler Construction*, Edinburgh, 1994.

[82] J. Hoogerbrugge, H. Corporaal, and H. Mulder. Software pipelining for transport-triggered architectures. In *Proc. 24th Annual Int. Workshop on Microprogramming*, pages 74–81, Albaquerque, New Mexico, November 1991.

[83] H. S. Hou.   A fast recursive algorithm for computing the discrete cosinge transform. *IEEE Trans. Acoustics, Speech, and Signal Processing*, 35(10):1455–1461, October 1987.

[84] X. Hu et al. Codesign of architectures for automotive powertrain modules. *IEEE Micro*, 14(4):17–25, August 1994.

[85] A.C. Hung. PVRG-JPEG codec 1.1. ftp:// havefun.stanford.edu/ pub/ jpeg/ JPEGDOCv1.1.tar.Z.

[86] A.C. Hung. PVRG-p64 codec 1.1. ftp:// havefun.stanford.edu/ pub/ p64/ P64DOCv1.1.tar.Z.

[87] A.C. Hung and T.H.-Y. Meng. A comparison of fast inverse discrete cosine transform algorithms. *Multimedia Systems*, 2(4):204–217, 1994.

[88] K.D. Huynh and T.M. Khoshgoftaar. A performance analysis of personal computers in a video conferencing environment. *Multimedia Systems*, 2(3):103–117, September 1994.

[89] C.T. Hwang, J.-H. Lee, and Y.-C. Hsu. A formal approach to the scheduling problem in high level synthesis. *IEEE Trans. CAD*, 10(4):464–475, April 1991.

[90] R. Babb II et al. Retargetable high performance fortran compiler challenges. In *COMPCON '93*, pages 137–146, 1993.

[91] Intel. *Intel Announces MMX Technology*. Intel, Intel Corporation, Mission College Blvd., Santa Clara, CA 95052-8119, USA, 1996. http:// www.intel.com/ pc-supp/ multimed/ mmx/.

[92] T. Ishiguro. VLSI in picture coding. *J. VLSI Signal Processing*, 5:115–120, 1993.

[93] T. B. Ismail, M. Abid, and A. Jerraya. COSMOS: A codesign approach for communicating systems. In *3rd Int. Workshop on Hardware/Software Codesign*, pages 17–24, Sept. 22-24, Grenoble, September 1994.

[94] ISO/IEC. *Informationtechnology – Coding of moving pictures and associated audio for digital storage media up to about 1.5 Mbit/s*, volume 11172. ISO/IEC, 1993.

[95] ISO/IEC. *Information technology - Digital compression and coding of continuous-tone still images, Requirements and Guidelines*, volume 10918-1. ISO/IEC, 1994.

[96] ITU-T. Series h recommendations. gopher:// info.itu.ch/ 11/ .1/ itudoc/ public/ gophertree/ .1/ .itu-t/ .rec/ .h.

[97] ITU-T. *Video Codec for Audiovisual Services at p × 64 kbit/s*. Recommendation H.261, Helsinki, March 1-12, March 1993.

[98] A. E. Jacquin. Image coding based on a fractal theory of iterated contractive image transformations. In *IEEE Trans. Image Processing*, pages 18–30, vol.1, no. 1, January 1992.

[99] A.E. Jacquin. Fractal image coding: A review. *Proc. IEEE*, 81(10):1451–1465, October 1993.

[100] R. Jain et al. Predicting system-level area and delay for pipelined and nonpipelined designs. *IEEE Trans. CAD*, 11(8):955–965, August 1992.

[101] N. Jayant, J. Johnston, and R. Safranek. Signal compression based on models of human perception. *Proc. IEEE*, 81(10):1385–1422, 1993.

[102] A. A. Jerraya et al. Linking system design tools and hardware design tools. In D. Agnew others, editor, *Computer Hardware Description Languages*. IFIP Trans., vol. A-32, 1993.

[103] M. Johnson. *Superscalar Microprocessor Design*. Prentice Hall, 1991.

[104] N. P. Jouppi and D. W. Wall. Available instruction-level parallelism for superscalar and superpipelined machines. In *Proc. 3rd Conf. Architectural Support for Programming Languages and Operating Systems*, pages 272–282, Boston, April 1989.

[105] JPEG. JPEG image compression: Frequently asked questions. http:// www.cis.ohio-state.edu/ hypertext/ faq/ usenet/ jpeg-faq/ faq.html.

[106] A. Kalavade and E. A. Lee. A hardware/software codesign methodology for DSP applications. *IEEE Design and Test*, 10(3):16–28, September 1993.

[107] A. Kalavade and E.A. Lee. A global criticality/local phase driven algorithm for the constrained hardware/software partitioning problem. In *3rd Int. Workshop on Hardware/Software Codesign*, pages 42–48, Sept. 22-24, Grenoble, September 1994.

[108] A. Kalavade and E.A. Lee. Manifestations of heterogeneity in hardware/software codesign. In *31st ACM/IEEE Design Automation Conference*, San Diego, CA, 1994.

[109] G. Kane and J. Heinrich. *MIPS RISC Architecture*. Prentice Hall, Englewood Cliffs, NJ, 1992.

[110] K. Kennedy, K.S. McKinley, and C.-W. Tseng. Analysis and transformation in an interactive parallel programming tool. *Concurrency: Practice and Experience*, 5(7):575–602, October 1993.

[111] K. Konstantinides and V. Bhaskaran. Monolithic architectures for image processing and compression. *IEEE Computer Graphics and Applications*, 12(6):75–86, November 1992.

[112] F. Kretz and F. Colaitis. Standardizing hypermedia information objects. In *IEEE Communications Magazine*, pages 60–70, May 1992.

[113] D. Kuck et al. The cedar system and an initial performance study. In *Proc. 20th Annual Int. Symp. Computer Architecture*, pages 213–223, San Diego, May 1993.

[114] E.D. Lagnese and D. E. Thomas. Architectural partitioning for system level synthesis of integrated circuits. *IEEE Trans. CAD*, 10(7):847–860, July 1991.

[115] A. Laine, editor. *Wavelet Theory and Application*. Kluwer Academic Publishers, Dordrecht, Netherlands, 1993.

[116] M.S. Lam and R.P. Wilson. Limits on control flow parallelism. In *19th Annual Int. Symp. Computer Architecture*, Gold Coast, Australian, 1992.

[117] M. Langevin, E. Cerny, J. Wilberg, and H.T. Vierhaus. Local microcode generation in system design. In *Code Generation for Embedded Processors*.

[118] M. Langevin, J. Wilberg, P. Plöger, and H.-T. Vierhaus. A codesign methodology for high performance embedded systems. In *High Performance Computing Symposium '95*, pages 353–364, Montreal, Canada, July 10 - July 12, July 1995.

[119] D. Lanneer, M. Cornero, G. Goossens, and H. De Man. Data routing: a paradigm for efficient data-path synthesis and code generation. In *7th Int. Workshop on High-Level Synthesis*, pages 17–21, Niagara-on-the-Lake, Ontario, Canada, May 18-20, May 1994.

[120] R. Lee. Precision architecture. *IEEE Computer*, 22(1):78–91, Jan 1989.

[121] R. Lee. Accelerating multimedia with enhanced microprocessors. *IEEE Micro*, 15(2):22–32, April 1995.

[122] R. L. Lee. Realtime MPEG video via software decompression on a PA-RISC processor. pages 186–192.

[123] D. B. Lenat. Artificial intelligence. *Scientific American*, 273(3):80–83, September 1995.

[124] C. Liem, T. May, and P. Paulin. *Instruction-Set Matching and Selection for DSP and ASIP Code Generation*. EDAC, 1994.

[125] M. Liou. Overview of the px64 kbps video coding standard. *Comm. ACM*, 34(4):59–63, April 1991.

[126] C. Liu, J. Peek, R. Jones, B. Buus, and A. Nye. *Managing Internet Information Services*. O'Reilly & Associates, Sebastopol, CA 95472, 1994.

[127] J. Madsen and J. P. Brage. Codesign analysis of a computer graphics application. *J. Design Automation for Embedded Systems*, 1(1/2):121–145, 1996.

[128] L. Maliniak. DAC focuses on synthesis and ESDA tool advances. *Electronic Design, May 30*, 42(11):51–66, May 1994.

[129] S. Mallat. A theory for multiresolution signal decomposition: The wavelet representation. *IEEE Trans. Pattern Anal. Mach. Intel.*, 11(7):674–693, July 1989.

[130] H.S. Malvar and D.H. Staelin. The LOT: Transform coding without blocking effects. *IEEE Trans. ASSP*, 37(4):553–559, April 1989.

[131] H. De Man. Design technology research for the ninties: More of the same? In *EuroDAC'92*, pages 592–596, 1992.

[132] P. Marwedel and W. Schenk. Cooperation of synthesis, retargetable code generation, and test generation in the MSS. In *Proc. European Design and Test Conf.*, pages 63–69, February 1993.

[133] M.C. McFarland, A. C. Parker, and R. Camposano. The high-level synthesis of digital systems. *Proc. IEEE*, 78(2):301–318, February 1990.

[134] A. Mendelsohn. Video as data: challenge at the leading edge. *Computer Design*, 33(7):96–102, June 1994.

[135] T. H. Meng, B. M. Gordon, E. K. Tsern, and A. C. Hung. Portable video-on-demand in wireless communication. *Proc. IEEE*, 83(4):659–680, April 1995.

[136] G. De Micheli. Computer-aided hardware-software codesign. *IEEE Micro*, 14(4):10–16, August 1994.

[137] T. Minami et al. A 300-MOPS video signal processor with a parallel architecture. *IEEE J. Solid-State Circuits*, 26(12):1868–1875, December 1991.

[138] M. Minsky and S. Papert. *Perceptrons*. MIT Press, Cambridge, MA, expanded edition edition, 1988.

[139] S. Mirapuri et al. The mips r4000 processor. In *IEEE Micro*, pages 10–22, April 1992.

[140] C. Monahan and F. Brewer. Symbolic modeling and evaluation of data paths. Technical Report ECE Technical Report #94-26, University of Santa Barbara, Department of Electrical Engineering, Santa Barbara, CA 93106-9560, October 1994.

[141] Morgan Kaufmann, San Francisco,. *The PowerPC Architecture*, 1994.

[142] MPEG. Moving picture expert group. http:// www.crs4.it/ HTML/ LUIGI/ MPEG/ mpegfaq.html.

[143] T. Murakami et al. A DSP architectural design of low bit-rate motion video codec. *IEEE Trans. Circuits and Systems*, 36(10):1267–1274, October 1989.

[144] H.-G. Musmann et al. Kompressionsalgorithmen für interaktive multimedia-systeme. *it+ti*, 35(2):4–18, April 1993.

[145] H. G. Musmann, P. Pirsch, and H.-J. Grallert. Advances in picture coding. *Proc. IEEE*, 73(4):523–548, April 1985.

[146] N. P. Negroponte. Products and services for computer networks. *Scientific American Special Issue: The Computer in the 21st Century*, pages 102–109, 1995.

[147] M. Ohta, M. Yano, and T. Nishitani. Wavelet picture coding with transform coding approach. In *IEICE Trans. Fundamentals*, pages 776–784, vol. E75-A, no. 7, July 1992.

[148] K. A. Olukotun et al. A software-hardware cosynthesis approach to digital system simulation. In *IEEE Micro*, pages 48–58, August 1994.

[149] K. Patel, B.C. Smith, and L.A. Rowe. Performance of a software MPEG video decoder. In *Proc. 1st ACM Int. Conf. on Multimedia*, Anaheim, CA, 1993.

[150] D. A. Patterson. Microprocessors in 2020. *Scientific American*, pages 48–51, September 1995.

[151] P.G. Paulin and J.P. Knight. Force-directed scheduling for the behavioral synthesis of ASIC's. *IEEE Trans. CAD*, 8(6):661–679, June 1989.

[152] P.G. Paulin, C. Liem, T.C. May, and S. Sutarwala. DSP tool requirements for embedded systems: A telecommunications industrial perspective. *J. VLSI Signal Processing*, 9:23–47, 1995.

[153] J. Peek, T. O'Reilly, and M. Loukides. *UNIX Power Tools*. O'Reilly & Associates, Sebastopol, CA, 1993.

[154] W. B. Pennebaker and J. L. Mitchell. *JPEG Still Image Data Compression Standard.* Van Nostrand Reinhold, New York, 1993.

[155] T. S. Perry. Technology 1994 - consumer electronics. In *IEEE Spectrum*, pages 30–34, January 1994.

[156] P. Pirsch. VLSI architectures for digital video signal processing. In P. Dewilde and J. Vandewalle, editors, *Computer Systems and Software Engineering*, pages 65–99. Kluwer Academic Publishers, Dordrecht, Netherlands, 1992.

[157] P. Pirsch, N. Demassieux, and W. Gehrke. VLSI architectures for video compression - a survey. *Proc. IEEE*, 83(2):220–246, February 1995.

[158] C.D. Polychronopoulos. Parallel programming issues. *Int J. High Speed Computing*, 5(3):413–473, 1993.

[159] T. Potter, M. Vaden, J. Young, and N. Ullah. Resolution of data and control-flow dependencies in the powerPC 601. *IEEE Micro*, 14(5):18–29, October 1994.

[160] J. Van Praet et al. Instruction set definition and instruction selection for ASIPs. In *7th Int. Symp. High-Level Synthesis*, Niagara-on-the-lake, Ontario, Canda, May 18-20, May 1994.

[161] L. Press. The internet and interactive television. In *ACM Communications*, pages 19–23, vol.36, no. 12, 140, December 1993.

[162] S. A. Przybylski. *Cache design: A performance-directed approach*. Morgan Kaufmann, San Mateo, 1990.

[163] S. C. Purcell and D. Galbi. C-Cube MPEG video processor. *Proc. SPIE Image Processing and Interchange*, 1659:30–40, 1992.

[164] S.C. Purcell and D. Galbi. The C-Cube CL550 JPEG image compression processor. In *Compcon Spring'91*, pages 318–323, Digest of Papers, 1991.

[165] J. M. Rabaey and M. Potkonjak. Estimating implementation bounds for real time DSP application specific circuits. *IEEE Trans. CAD*, 13(6):669–683, 1994.

[166] M. Rabbani. Selected papers on image coding and compression. In *Proc. SPIE*, vol. MS48, Milestone Series, 1992.

[167] K.R. Rao. *Discrete Cosine Transform*. Academic Press, New York, 1990.

[168] B. R. Rau and J. A. Fisher. Instruction-level parallel processing: History, overview, and perspective. In *J. Supercomputing*, pages 9–50, vol.7, no. 2, May 1993.

[169] W. Rosenstiel and H. Krämer. Scheduling and assignment in high level synthesis. In R. Camposano and W. Wolf, editors, *High-Level VLSI Synthesis*, pages 355–382. Kluwer Academic Publishers, Boston, 1991.

[170] P. A. Ruetz et al. A high-performance full-motion video compression chip set. *IEEE Trans. Circuits And Systems on Video Technology*, 2(2):111–122, 1992.

[171] B. Ryan. Alpha rides high. *BYTE*, 19(10):197–198, October 1994.

[172] R. Schäfer and T. Sikora. Digital video coding standards and their role in video communications. *Proc. IEEE*, 83(6):907–924, June 1995.

[173] S. E. Schulz. Behavioral synthesis: Concept to silicon. In *ASIC&EDA*, pages 12–26, August 1994.

[174] S. Segars, K. Clarke, and L. Goudge. Embedded control problems, thumb, and the ARM7TDMI. *IEEE Micro*, 15(5):22–30, October 1995.

[175] A. Sharma and R. Jain. Estimating architectural resources and performance for high-level synthesis applications. In *30th ACM/IEEE Design Automation Conference*, pages 355–360, 1993.

[176] Y.-H. Shiau and C.-P. Chung. Adoptability and effectiveness of microcode compaction algorithms in superscalar processing. *Parallel Computing*, 18(5):497–510, 1992.

[177] R. L. Sites. Alpha AXP architecture. In *Comm ACM*, pages 33–44, vol.36, no. 2, February 1993.

[178] A. Smailagic and D.P. Siewiorek. The VuMan 2 wearable computer. *IEEE Design & Test*, 10(3):56–67, September 1993.

[179] J. E. Smith and G. S. Sohi. The microarchtecure of superscalar processors. *IEEE Proc.*, 83(12):1609–1624, December 1995.

[180] M.D. Smith. Tracing with pixie. Report, CSL-TR-91-497, Computer System Lab., November 1991.

[181] R. M. Stallman and R. McGrath. *GNU Make*. Free Software Foundation, Cambridge, MA, 0.27 beta edition, May 1990.

[182] P. Stravers. *Embedded System Design*. PhD thesis, Delft Technical University, Netherlands, 1994.

[183] P. Stravers and E. Aardoom. A processor framework customized for navigation computations. *Integration, the VLSI J.*, 14:197–214, 1992.

[184] B. Stroustrup. *The C++ Programming Language*. Addison-Wesley, Reading, MA, 1991.

[185] Sun. Writing java programs. http: //java.sun.com /doc /programmer.html.

[186] Sun. *SPARC Architecture Manual*. Sun Microsystems, Mountain View, CA, 1988.

[187] I. Tamitani et al. A real-time video signal processor suitable for motion picture coding applications. *IEEE Trans. Circuits and Systems*, 36(10):1259–1266, October 1989.

[188] S.-I. Uramoto et al. A 100-MHz 2-D discrete cosine transform core processor. *IEEE J. Solid-State Circuits*, 27(4):492–498, April 1992.

[189] R. Vetter, C. Spell, and C. Ward. Mosaic and the worl-wide web. *IEEE Computer*, 27(10):49–57, October 1994.

[190] W3. Hypertext markup language (HTML). http:// www.w3.org / hypertext / WWW / MarkUp / MarkUp.html.

[191] R. A. Walker and R. Camposano. *A Survey of High-Level Synthesis Systems*. Kluwer Academic Publishers, Boston, MA, 1991.

[192] G. K. Wallace. The JPEG still picture compression standard. *IEEE Trans. Consumer Electronics*, 38(1):18–34, February 1992.

[193] P. Wayner. Digital video goes real-time. In *BYTE*, pages 107–112, January 1994.

[194] P. Wayner. SPARC strikes back. *BYTE*, 19(11):105–112, November 1994.

[195] P. Wayner. VLIW question. *BYTE*, 19(11):287–288, November 1994.

[196] N. Weste. OK, if these CAD tools are so great, why isn't my chip design on schedule? In *Int. Conf. Computer Design*, pages 2–8, Cambridge, MA, Oct. 10-12, October 1994.

[197] J. Wilberg. Development of a processor architecture for h.261-codecs. Master's thesis, University of Hannover, Hannover, Germany, November 1992.

[198] J. Wilberg. Codesign for real-time video compression – additional information. http:// borneo.gmd.de/ ~wilberg/ desktop/ phdBook/ README.html, 1997.

[199] J. Wilberg, R. Camposano, M. Langevin, P. Plöger, and T. Vierhaus. Cosynthesis in CASTLE. In G. Saucier and A. Mignotte, editors, *Novel Approaches in Logic and Architecture Synthesis*, pages 355–366. Chapman & Hall, London, 1995.

[200] J. Wilberg, R. Camposano, and W. Rosenstiel. Design flow for hardware/software cosynthesis of a video compression system. In *3rd Int. Workshop on Hardware/Software Codesign*, pages 73–80, Sept. 22-23, Grenoble, September 1994.

[201] J. Wilberg, R. Camposano, U. Westerholz, and U. Steinhausen. Design of an embedded video compression system - a quantitative approach. In

Int. Conf. Computer Design, pages 428–431, Cambridge, MA, Oct. 10-12, October 1994.

[202] J. Wilberg et al. Hierarchical multiprocessor system for video signal processing. *Proc. SPIE*, 1818, November 1992.

[203] J. Wilberg, A. Kuth, W. Rosenstiel, and H.-T. Vierhaus. A design management system for the WWW. In *8. E.I.S. Workshop, University of Hamburg, Germany*, D-53754 Sankt Augustin, Germany, 8. and 9. April 1997 1997. GMD.

[204] J. Wilberg, P. Plöger, R.Camposano, M. Langevin, and T. Vierhaus. Codesign of hardware, software, and algorithms - a case study. In *Proceedings International Symposium on Circuits and Systems*, volume 4, pages 552–555, Atlanta, May12–15 1996. IEEE.

[205] R. Wilson et al. The SUIF compiler system. http:// suif.stanford.edu/ suif/ suif-overview/ suif-overview.html, 1994.

[206] N. Wirth. A plea for lean software. *IEEE Computer*, 28(2):64–68, February 1995.

[207] I.H. Witten, R.M. Neal, and J.G. Cleary. Arithmetic coding for data compression. *Communications of the ACM*, 30(6):520–540, June 1987.

[208] W. Wolf. Hardware-software co-design of embedded systems. *Proc. IEEE*, 82(7):967–989, July 1994.

[209] W. Wolf et al. Tigerswitch: A case study in embedded computing system design. In *3rd Int. Workshop on Hardware/Software Codesign*, pages 89–96, Sept. 22-23, Grenoble, September 1994.

[210] M. Wolfe. *Optimizing Supercompilers for Supercomputers*. MIT Press, Cambridge, 1989.

[211] J.W. Woods and S.D. O'Neil. Subband coding of images. *IEEE Trans. ASSP*, 34(5):1278–1288, October 1986.

[212] H. Yamauchi et al. Architecture and implementation of a highly parallel single-chip video DSP. *IEEE Trans. Circuits And Systems For Videotechnologie*, 2(2):207–220, June 1992.

[213] J. C.-Y. Yang, G. De Micheli, and M. Damiani. *Scheduling with Environmental Constraints based on Automata Representations*. EDAC, 1994.

[214] R.W. Young and N.G. Kingsbury. Video compression using lapped transform motion/estimation/compensation and coding. *Proc. SPIE Visal Communications and Image Processing*, 1818:276–288, 1992.

[215] C.-G. Zhou, L. Kohn, D. Rice, I. Kabir, A. Jabbi, and X.-P. Hu. MPEG video decoding with the UltraSPARC visual instruction set. pages 470–475.

Index